Nanofiltration Membranes

Synthesis, Characterization, and Applications

Nanofiltration Membranes

Synthesis, Characterization, and Applications

Lau Woei Jye
Ahmad Fauzi Ismail

CRC Press is an imprint of the
Taylor & Francis Group, an **informa** business

CRC Press
Taylor & Francis Group
6000 Broken Sound Parkway NW, Suite 300
Boca Raton, FL 33487-2742

First issued in paperback 2019

ISBN-13: 978-1-4987-5137-7 (hbk)
ISBN-13: 978-0-367-87577-0 (pbk)

Visit the Taylor & Francis Web site at
http://www.taylorandfrancis.com

and the CRC Press Web site at
http://www.crcpress.com

Contents

Preface ... vii
Acknowledgments ... ix
Authors ... xi
Abbreviations ... xiii

1. Introduction ... 1
1.1 History of Nanofiltration Membrane and Applications ... 1
1.2 Global Market and Research Trend of NF ... 4
1.3 Transport Mechanisms of NF ... 7
1.3.1 Irreversible Thermodynamics ... 8
1.3.2 Steric Hindrance Pore Model and Hagen–Poiseuille Model ... 9
1.3.3 Teorell–Meyer–Sievers Model ... 11
1.3.4 Donnan Steric Pore-Flow Model ... 12
References ... 13

2. Synthesis of Nanofiltration Membrane ... 15
2.1 Overview of NF Membrane Synthesis ... 15
2.2 PA Thin Film (Nano)Composite Membrane via IP Technique ... 15
2.3 Asymmetric Membrane via Phase Inversion Technique ... 20
2.4 Multilayer Membrane via Layer-by-Layer Assembly Method ... 27
References ... 30

3. Advanced Materials in Nanofiltration Membrane ... 33
3.1 Overview of Advanced Materials Used in Thin Film (Nano) Composite Membrane ... 33
3.2 Advanced Materials in PA Thin Layer ... 34
3.2.1 Monomer ... 34
3.2.2 Surfactant/Additive ... 47
3.2.3 Inorganic Nanomaterial ... 50
3.3 Advanced Materials in Microporous Substrate ... 60
3.3.1 Polymer/Polymer–Polymer-Based Substrate ... 60
3.3.2 Polymer/Inorganic Nanocomposite Substrate ... 70
References ... 72

4. Technical Challenges and Approaches in Nanofiltration Membrane Fabrication ... 79
4.1 TFC Hollow Fiber Membrane ... 79
4.1.1 Challenges ... 79
4.1.2 Innovative Approaches ... 80

4.2 TFN Flat Sheet Membrane 81
4.2.1 Challenges 81
4.2.2 Innovative Approaches 83
4.2.2.1 Surface Modification of Nanomaterials 83
4.2.2.2 Use of Metal Alkoxides 86
4.2.2.3 Modified/Novel IP Techniques 87
4.2.2.4 Alignment of Nanotubes/Fillers 89
References 91

5. Characterization of Nanofiltration Membrane 95
5.1 Overview of NF Characterization 95
5.2 Instruments/Methods for Chemical Properties Assessment 96
5.2.1 Attenuated Total Reflectance-FTIR Spectroscopy 96
5.2.2 Zeta Potential 99
5.2.3 X-Ray Photoelectron Spectroscopy 101
5.2.4 X-Ray Diffractometry 107
5.2.5 Nuclear Magnetic Resonance Spectroscopy 109
5.3 Methods for Physical Properties Assessment 110
5.3.1 Electron Microscopy 110
5.3.2 Atomic Force Microscopy 113
5.3.3 Positron Annihilation Spectroscopy 114
5.3.4 Surface Contact Angle 119
5.4 Characterization on Membrane Permeability and Selectivity ... 121
5.4.1 Permeability 121
5.4.2 Selectivity 123
5.4.2.1 Neutral Solutes 123
5.4.2.2 Charged Solutes 126
5.4.2.3 Dyes 127
5.5 Characterization on Membrane Stabilities 129
5.5.1 Chlorination 129
5.5.2 Thermal 130
5.5.3 Solvent 131
5.5.4 Filtration 132
References 133

6. Applications of Nanofiltration Membrane 139
6.1 Overview of NF Membrane Applications 139
6.2 NF of Aqueous Solvent Systems 139
6.2.1 Water Treatment 140
6.2.2 Wastewater Treatment 144
6.3 NF of Nonaqueous Solvent Systems 149
6.3.1 Solvent Recovery 149
6.3.2 Recovery/Purification of Pharmaceutically Active Ingredients and Valuable Catalysts 152
References 157

Index 161

Preface

Membrane science and technology is an expanding field and has become a prominent part of many processes in industry. Books on the science and technology of membranes made of polymeric materials are abundant in the market. But there have been only a few books available today that exclusively address polymeric nanofiltration (NF) membranes. One of the popular NF books is the one edited by A.G. Fane, A. Schaefer, and T. David Waite. This book entitled *Nanofiltration: Principles and Applications* was published by Elsevier in 2005.

This new book intends to provide a comprehensive overview of the development of NF membrane technology for the past 10 years. It can be considered as complementary and updated content to the existing NF books. What makes this new book different from other books published is its strong emphasis on the fabrication techniques that one can employ to fabricate NF membranes of different morphologies. The potential of advanced materials in improving the properties of NF membrane is also covered in this book and is discussed according to several important areas that are divided into monomer, surfactant/additive, inorganic nanomaterials, and microporous substrates. In certain cases, the membranes made of newly synthesized materials demonstrate not only excellent antifouling resistance and/or greater antibacterial effect, but also possibly overcome the trade-off effect between water permeability and solute selectivity. Comments are also provided on some experimental results that might be less convincing from a technical point of view. In addition, technical challenges in fabricating the new generation of NF membranes are also reviewed and the possible approaches to overcome the challenges are provided in the hope that it will assist researchers to produce an NF membrane with a defect-free surface.

This book also discusses the characterization methods used in assessing the physical and chemical properties of membranes as well as separation characteristics and performance stability. The content of this section is of importance to help new researchers in the field of NF to quickly learn various NF characterization methods without studying a large number of different reference books. It is also the aim of this book to give readers insights into the experimental planning and subsequent interpretation of the characterization techniques. This book concludes with relevant case studies on the use of NF membranes in industrial implementations of both aqueous and nonaqueous media.

We are deeply indebted to many of our colleagues and students of the Advanced Membrane Technology Research Centre of the Universiti Teknologi Malaysia for their assistance during the writing of this book. A few whose names stand out are Dr. Ong Chi Siang, Dr. Nurasyikin Misdan, Dr. Daryoush

Emadzadeh, Dr. Nur Aimie Abdullah Sani, Dr. Mohammed Ghanbari, Dr. Goh Pei Sean, Dr. Rasoul Jamshidi Gohari, Mr Lai Gwo Sung, Mr Ihsan Wan Azelee, and Mr Ng Be Cheer. We also gratefully thank Professor Emeritus Takeshi Matsuura of the University of Ottawa, Canada for his detailed and helpful review of the manuscript.

Dr. Lau Woei Jye
Dr. Ahmad Fauzi Ismail

Acknowledgments

Dr. Woei Jye wishes to express his gratitude to his parents, wife, daughter, and three lovely sisters for their endless love, support, and encouragement during the writing of this book. He also thank his former students who were involved in the research and development of nanofiltration membranes, for their valuable contributions.

Dr. Ismail wishes to express his gratitude to family for their continuous support and encouragement during the writing of this book. He also thank all his staff and students for their hard work, dedication and support during their precious years at AMTEC.

Authors

Lau Woei Jye was born in Johor, Malaysia, in 1983 and earned a bachelor of engineering in chemical-gas engineering (2006) and a PhD in chemical engineering (2009) at the Universiti Teknologi Malaysia (UTM), Malaysia. Currently, he is a senior lecturer at the Faculty of Chemical and Energy Engineering (FCEE) and a research fellow at the Advanced Membrane Technology Research Centre (AMTEC), UTM. Before working at UTM, he was an assistant professor at Universiti Tunku Abdul Rahman (UTAR), Kuala Lumpur. Dr. Woei Jye has strong research interests in the field of water and wastewater treatment processes using membrane-based technology. He has published more than 80 scientific papers, 10 reviews, one book, and seven book chapters. He has supervised nine PhD and two master's students to successful completion, with six further postgraduate students currently under his sole or co-supervision. He was the recipient of the *Australian Endeavour Research Fellowship 2015 and UI-RESOLV Program 2016.*

Ahmad Fauzi Ismail is the Founding Director of Advanced Membrane Technology Research Center (AMTEC), Universiti Teknologi Malaysia (UTM). He earned a bachelor of engineering in petroleum engineering (1989) and a master's in chemical engineering (1992) at the UTM, Malaysia. He earned a PhD in 1997 from University of Strathclyde, UK. He has more than 20 years of experience in the development of membrane technology for various applications. He has published more than 450 scientific papers, 6 books, 3 edited books and more than 40 book chapters. He has won many outstanding awards, such as the Merdeka Award 2014, IChemE Malaysia Inventor of the Year Awards 2014, and the Malaysian Toray Science and Technology Award 2014. He is a Fellow of The Academy of Sciences Malaysia, Chartered Engineer in the UK (CEng) and a Fellow of the Institution of Chemical Engineers (FIChemE).

Abbreviations

The following lists key abbreviations used in this book.

2,2′-OEL	2,2′-Oxybis-ethylamine
6FAPBS	2,5-Bis(4-amino-2-trifluoromethylphenoxy) Benzenesulfonic acid
6FBABDS	4,4′-Bis(4-amino-2-trifluoromethylphenoxy) Biphenyl-4,4′-disulfonic acid
AAPTS	*N*-[3-(trimethoxysilyl)propyl]ethylenediamine
AFM	Atomic force microscopy
APDEMS	3-Aminopropyldiethoxymethylsilane
API	Active pharmaceutical ingredient
APS	3-Aminopropyltriethoxysilane
ATR	Attenuated total reflectance
β-CD	β-Cyclodextrin
BOD	Biological oxygen demand
BPA	Bisphenol A
BSA	Bovine serum albumin
BTEC	Biphenyl tetraacyl chloride
CA	Cellulose acetate
CAGR	Compound annual growth rate
CAIP	Co-solvent assisted interfacial polymerization
CC	Cyanuric chloride
CHMA	1,3-Cyclohexanebis(methylamine)
CL	Cross-linking
CMCNa	Sodium carboxymethyl cellulose
CN	Cellulose nanofiber
CNT	Carbon nanotube
COD	Chemical oxygen demand
CTAB	Cetyl trimethyl ammonium bromide
Da	Dalton
DAP	Diaminopiperazine
DAPA	3,5-Diamino-N-(4-aminophenyl)benzamide
DAPP	1,4-Bis(3-aminopropyl)piperazine
DBES	Doppler broadening energy spectra
DETA	Diethylenetriamine
DI	Deionized
DMF	Dimenthylformanide
DMSO	Dimenthylsulfoxide
DNA	Deoxyribonucleic acid
DSPM	Donnan steric pore-flow model

EC	Emerging contaminant
EDC	Endocrine disrupting chemical
EGME	Ethylene glycol monomethyl ether
FESEM	Field emission scanning electron microscopy
FFA	Free fatty acid
FO	Forward osmosis
FTIR	Fourier transform infrared
GO	Graphene oxide
GTIs	Genotoxic impurities
HFA	Hexafluoroalcohol
HMPA	Hexamethyl phosphoramide
HNT	Halloysite nanotube
H-OMC	Hydrophilized mesoporous carbon
HPEI	Hyperbranched polyethyleneimine
HTC	1,3,5-Tricarbonyl chloride
ICIC	5-Isocyanato-isophthaloyl chloride
ICP	Internal concentration polarization
IEP	Isoelectric point
IP	Interfacial polymerization
IPC	Isophthaloyl chloride
LBL	Layer-by-layer
LSMM	Hydrophilic surface modifying macromolecule
MBR	Membrane bioreactor
MEK	Methyl ethyl ketone
MEOA	Monoethanolamine
MF	Microfiltration
MMPD	*m*-Phenylenediamine-4-methyl
MOF	Metal-organic framework
MPD	*m*-Phenyldiamine
MSN	Mesoporous silica nanoparticle
MW	Molecular weight
MWCNT	Multi-walled carbon nanotube
MWCO	Molecular weight cut-off
NF	Nanofiltration
NMP	*n*-Methyl-2-pyrrolidone
NMR	Nuclear magnetic resonance
NTSC	1,3,6-Trisulfonylchloride
OCM	Ordered mesoporous carbon
PA	Polyamide
PAA	Polyacrylic acid
PAH	Poly(allyl amine) hydrochloride
PALS	Positron annihilation lifetime spectroscopy
PAN	Polyacrylonitrile
PANI	Polyaniline
PAS	Positron annihilation spectroscopy

PBI	Polybenzimidazole
PD	Polydopamine
PDADMAC	Poly(diallyl-dimethylammonium chloride)
PDMS	Polydimethlysiloxane
PE	Polyelectrolyte
PEEK	Poly(ether ether ketone)
PEG	Polyethylene glycol
PEI	Polyetherimide
PES	Polyethersulfone
PhAC	Pharmaceutically active compound
PhTES	Phenyl triethoxysilane
PI	Polyimide
PIP	Piperazine
PMMA	Polymethyl methacrylate
POP	Persistent organic pollutant
PP	Polypropylene
PPBES	Copoly(phthalazinone biphenyl ether sulfone)
PPD	*p*-Phenylenediamine
PPEA	Poly(phthalazinone ether amide)
PPENK	Poly(phthalazinone ether nitrile ketone)
PPESK	Poly(phthalazinone ether sulfone ketone)
PPSU	Polyphenylsulfone
PPy	Polypyrrole
PS	Polystyrene
PSD	Pore size distribution
PSF	Polysulfone
PSS	Poly(styrene sulfonate)
PTFE	Poly(tetrafluoroethylene)
PTSC	Polythiosemicarbazide
PVA	Poly(vinyl alcohol)
PVAm	Poly(vinylamine)
PVDF	Polyvinylidene difluoride
PVP	Polyvinylpyrrolidone
PVS	Poly(vinyl sulfate potassium salt)
PWF	Pure water flux
PWP	Pure water permeability
RMS	Root mean square
RNA	Ribonucleic acid
RO	Reverse osmosis
SAD	Surface area difference
SDI	Silt density index
SDS	Sodium dodecyl sulfate
SEM	Scanning electron microscopy
SHP	Steric hindrance pore
SLS	Sodium lauryl sulfate

SPEEK	Sulfonated poly(ether ether) ketone
SPESS	Sulfonated poly(ether sulphide sulfone)
SPPESK	Sulfonated poly(phthalazinone ether sulfone ketone)
SRNF	Solvent resistant nanofiltration
SRU	Sulfate removal unit
SWNT	Single-walled nanotube
TBP	Tributyl phosphate
TCI	Thionyl chloride
TDS	Total dissolved solid
TEBAB	Triethyl benzyl ammonium bromide
TEBAC	Triethyl benzyl ammonium chloride
TEM	Transmission electron microscopy
TEOA	Triethanolamine
TEPA	Tetraethylenepentamine
TETA	Triethylenetetramine
TFC	Thin film composite
TFN	Thin film nanocomposite
THF	Tetrahydrofuran
TMBAB	Trimethyl benzyl ammonium bromide
TMC	Trimesoyl chloride
TOC	Total organic carbon
TPP	Triphenyl phosphate
TTIP	Tetra isopropoxide
UF	Ultrafiltration
VACNT	Vertically aligned carbon nanotube
XPS	X-ray photoelectron spectroscope
XRD	X-ray diffraction

1

Introduction

1.1 History of Nanofiltration Membrane and Applications

Nanofiltration (NF) is the fourth class of pressure-driven membranes born after microfiltration (MF), ultrafiltration (UF), and reverse osmosis (RO). It was first developed in the late 1970s as a variant of RO membrane with reduced separation efficiency for smaller and less charged ions such as sodium and chloride. As the term NF was not known in the 1970s, such membrane was initially categorized as either loose/open RO, intermediate RO/UF, or tight UF membrane. The term NF appears to have been first used commercially by the FilmTec Corporation (now The Dow Chemical Company) in the mid-1980s to describe a new line of membrane products having properties between UF and RO membrane. Owing to the uniqueness and meaningfulness of the word NF, other membrane scientists have begun using it. The widespread use of this word today is testament to the need for just such a descriptor in the membrane lexicon (Schafer et al., 2005).

Figure 1.1 shows the distinction between the pressure-driven membrane processes with respect to solute rejection capability. NF membrane that is positioned in the lower end of UF and upper end of RO has a molecular weight cutoff (MWCO) of about 200–1000 Dalton (Da), which corresponds to pore sizes between 0.5 and 2 nm. Unlike UF and RO membranes which generally carry no charge on their surface, NF membrane often carries positive or negative electrical charges (Strathmann, 2011). In most cases, NF membranes are negatively charged in neutral or alkaline conditions and positively charged in highly acidic condition. In view of this, the separation of NF membrane is governed by three distinct mechanisms, namely the steric hindrance (or size sieving), electrostatic (Donnan) exclusion, and dielectric exclusion.

The first generation of NF membranes in the early 1970s was made of cellulose acetate (CA) or its derivatives. These membranes were synthesized based on the well-known Loeb–Sourirajan's dry–wet phase inversion technique in which a homogenous polymeric solution was cast on a glass plate followed by partial evaporation of solvent before immersing in a bath

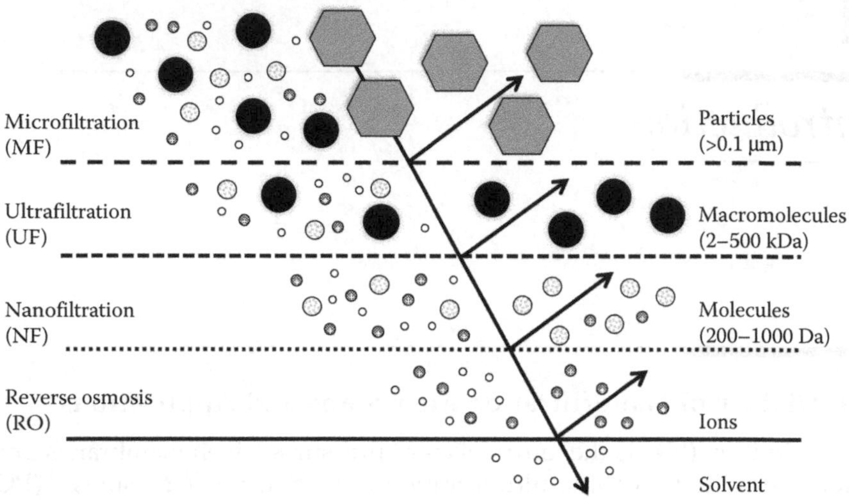

FIGURE 1.1
Separation spectrum of membrane.

containing a non-solvent coagulant. Originally, this phase inversion technique was used to produce a dense membrane structure for RO application, but it was realized that by varying the condition of post treatment (heat treatment), a membrane with relatively larger pores could be tailored and used for specific applications, particularly those requiring a lower degree of NaCl rejection and lower operating pressure (Loeb and Sourirajan, 1964; Cohen and Loeb, 1977). Nevertheless, it soon became evident that the poor biological and chemical stability of cellulose-based membranes have limited the range of industrial applications as these membranes always suffered from continual changes in water flux and solute rejection during operation.

Because of these reasons, the second generation of NF membrane based on noncellulosic materials was developed. This membrane is a thin film composite (TFC) membrane consisting of three different layers: an ultrathin polyamide (PA) selective layer on the top surface, a microporous interlayer, and a nonwoven polyester bottom layer, as illustrated in Figure 1.2. The advantage of a TFC membrane is that each layer can be independently optimized to achieve desirable membrane performance with respect to water permeability, solute selectivity, and chemical and thermal stability (Lau et al., 2012). Compared to the asymmetric CA membrane which can be synthesized via a single-step fabrication process, the PA composite membrane is typically fabricated through a two-step process. The top PA thin selective layer is synthesized by interfacial polymerization (IP) of amine monomer (e.g., *m*-phenylenediamine [MPD]) with acyl chloride monomer (e.g., trimesoyl chloride [TMC]) while the microporous interlayer is fabricated via the

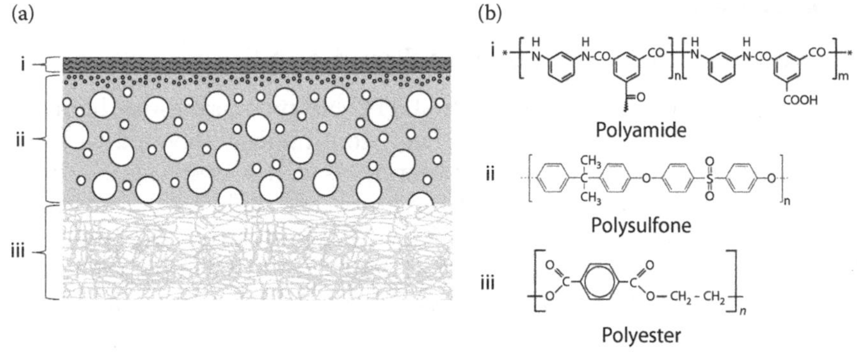

FIGURE 1.2
(a) Illustration of typical structure of TFC NF membrane consisting of (i) top selective layer, (ii) microporous substrate, and (iii) nonwoven fabric and (b) structure of polymers that are commonly used for each layer.

dry–wet phase inversion technique. Even though the TFC membrane was reported in the literature in the 1970s, commercial TFC membranes for NF application were only available in the second half of the 1980s after years of research and development. Currently, the TFC membrane is regarded as the most popular and reliable material in the membrane market. Permeate flow rate and its quality have been improved 10 times more than that at the beginning (Li et al., 2008).

The earliest documented NF application was a potable water application in Florida in the late 1970s (Paulson, 2015). It was probably the first commercial and intentional use of NF membranes (made of CA) for removing color molecules while allowing monovalent ions like sodium to pass through. This membrane partially demineralized the feed water, removing 10%–90% of dissolved solids, compared with up to 99.5% for typical RO. Low-level calcium hardness remaining in the NF permeate imparts a sweet taste to the water. Statistics revealed that NF had produced an estimated 1.4 million m^3 day^{-1} of drinking water for US communities in 2007 (Hanft, 2010). This figure was significantly higher compared to the total volume recorded in 1996, that is, approximately 0.23 million m^3 day^{-1}. Other application of NF in the early days was found in the purification step of a critical manufacturing process in which turbular NF was applied for desalination of dyes and brighteners (*Filtration+Separation*, 2005). Starting in the late 1970s, NF processes have gradually found their way into a wide range of industrial applications, serving a viable alternative to conventionally used separation processes.

Although NF membrane separation technology is presently being used in a variety of applications and has generated businesses totaling over a billion US dollars annually, the search has always been on for better membranes by taking into consideration other aspects such as economic competitiveness compared to other alternative separation methods.

1.2 Global Market and Research Trend of NF

Since the commercialization of the first TFC NF membranes in the market in the 1980s, considerable effort has been continuously devoted to further improve the properties of synthesis materials as well as synthesis conditions. This has led to the production of NF membranes with various separation characteristics, allowing applications for various industrial processes.

Table 1.1 summarizes the current major manufacturers of NF membranes for both aqueous and nonaqueous applications. The membrane products of

TABLE 1.1

Current Major Manufacturers of NF Membranes for Aqueous and Nonaqueous Applications

Medium	Membrane Manufacturer	NF Membrane (MWCO)[a]
Aqueous	The Dow Chemical Company	NF40 (180)
		NF90 (200)
		NF200 (300)
		NF270 (200–300)
	GE Osmonics	Desal 5 DK (150–300)
		Desal 5 DL (150–300)
	Nitto Denko	NTR-729 HF (700)
		NTR 7450 (600–800)
	Toray Industries	UTC20 (180)
		UTC60 (150)
	Pentair X-Flow	HFW1000 (1000)
	Hoechst CA	CA30 (1000)
	Koch Membrane Systems	SR3D (200)
		MPF34 (200)
		MPF36 (1000)
Organic[b]	SolSep BV	NF010206 (300–500)
		NF030705 (500)
		NF030306 (500–1000)
	Koch Membrane Systems	MPF44 (250)
		MPF50 (700)
	Evonik-MET Ltd.	DuraMem® (150–900)
		PuraMem® (280–600)
	AMS Technologies Membranes	SX-3014 (~330)
		SX-3016 (~350)
	Membrane Extraction Technology	STARMEM™ 120 (200)
		STARMEM™ 122 (220)
		STARMEM™ 240 (400)

[a] Estimated MWCO (Dalton) is shown in bracket.

[b] The MWCO of membrane might vary depending on the type of organic solvent and solute tested.

each manufacturer are also provided together with their estimated MWCO. The Dow Chemical Company and GE Osmonics are among the leading players in the global NF membrane technology marketplace. Other major players include Nitto Denko, Toray Industries, Koch Membrane Systems, Evonik-MET Ltd., and Membrane Extraction Technology.

A recent market analyst report entitled *"Global Markets and Technologies for Nanofiltration"* revealed that global market for NF membrane and its process is projected to grow to US$ 445.1 million by 2019 with a 5-year compound annual growth rate (CAGR) of greater than 15.5% (BCC Research, 2014). Of the total market share projected, the NF for the water and wastewater treatment segment is expected to contribute about two-thirds or approximately US$ 335 million. This value is significantly higher than the market value recorded in 2007, that is, US$ 70.9 million (BCC Research, 2007). The future for NF expansion in the sector of water and wastewater treatment is generally perceived to be optimistic. The level of optimism is reinforced by an understanding of the key factors driving the NF market such as stricter environmental regulations and the growing confidence in the performance of nanotechnology.

In addition to water and wastewater treatment, NF has also shown significant penetration into other industrial sectors such as food and beverage processing, pharmaceutical and biomedical processes, chemical and petrochemical applications, etc. Although these sectors only contribute a small portion to the total market share, their growth is relatively fast compared to the water and wastewater segment. Pharmaceutical and biomedical processes, for instance, is the fastest moving segment in the overall market with a CAGR of about 20%. With this growing rate, the segment is projected to reach US$ 22.7 million by 2019 (BCC Research, 2014). NF membrane for this sector is mainly used for plasma purification, protein fractionation, preparation of desalted and concentrated antibiotics, filtration of deoxyribonucleic acid (DNA), ribonucleic acid (RNA), and endotoxins, etc. Other applications of NF for various sectors are summarized in Table 1.2. With more and more applications proving to be successful, industrialists today feel more confident about what can be expected from an NF membrane.

With respect to worldwide research and development activities, the growth of NF research publications has increased tremendously since 1996. Figure 1.3 presents the number of published papers in journals per year in the NF field. About 7000 papers related to the NF membrane were documented over the recent past 20 years (1996–2015), with most of these studies published in international peer-reviewed journals, to name a few, *Desalination* (13.2%), *Journal of Membrane Science* (13.0%), *Separation and Purification Technology* (4.1%), *Desalination and Water Treatment* (3.5%) *and Water Research* (2.0%). Of the total number of papers published, about 43% was recorded for the years between 2011 and 2015. It can be said that NF membrane and its applications are still a hot field of study in today's world and are expected to continue growing for years to come.

TABLE 1.2
Commercial Applications of NF Membranes for Aqueous and Nonaqueous Process

Sector	Applications	Solvent Medium
Water treatment	Water softening, color removal, micropollutants elimination, pretreatment to RO process	Aqueous
Wastewater treatment	Leachate wastewater, textile effluent, emerging contaminants (ECs) removal, effluent from pulp and paper process	Aqueous
Food and dairy	Whey pre-concentration, whey protein desalination, caustic and acid recovery, gelation concentration	Aqueous and nonaqueous
Pharmaceutical and biomedical	Fractionation of proteins, plasma purification, filtration of DNA, RNA, and endotoxins, preparation of desalted and concentrated antibiotics, recovery of 6-amino penicillanic acid from waste stream	Aqueous and nonaqueous
Oil and gas	Solvent recovery from lube oil and hydrocarbon solvent mixtures, removal of sulfate from seawater before offshore reservoir injection	Aqueous and nonaqueous

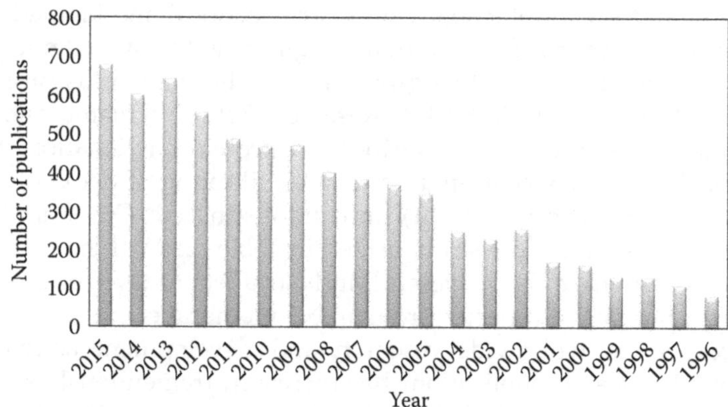

FIGURE 1.3
Growth of research NF activity up to year 2015 represented as a plot of number of publications in refereed journals for each year. (Adapted from Scopus, Date of access: March 24, 2016.)

Figure 1.4 shows the published papers of the top 10 most productive countries in the field of NF over the period of 1996–2015. The United States and China were the two leading countries with 1078 and 990 papers published, respectively. This is considerably more than the next two countries in the top 10 list, France and Germany, who each produced a respectable 524 and

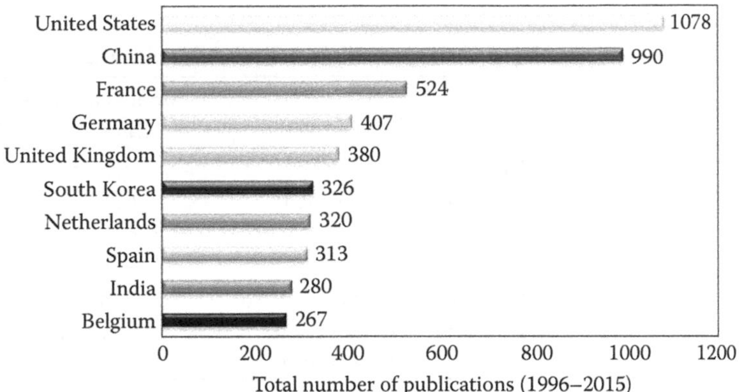

FIGURE 1.4
Top 10 countries by the total number of NF scientific papers published in the period of 1996–2015. (Adapted from Scopus, Date of access: March 24, 2016.)

407 scientific papers, respectively. It is also worth mentioning that the top 10 countries contributed to more than two-thirds of the total number of papers published worldwide. Most of the considered publications in Figures 1.3 and 1.4 are concerned with the synthesis and characterization of NF membrane and experimental studies on the effects of the operating/process conditions. Other papers reported in the journals are focused on the performance evaluation of NF hybrid process and theoretical models.

1.3 Transport Mechanisms of NF

To describe and possibly predict solvent fluxes and solute rejections through NF membranes, the transport mechanism of both solvents and solutes has to be understood. Three kinds of models have been used to describe the transport through NF membranes. The first group is based on irreversible thermodynamics and considers the membrane as a "black box" separating two phases far from equilibrium, without taking into account the membrane property. The other two groups describe the solute transport as a function of the structural and physiochemical parameters of membranes. Understanding the transport mechanism is important for selecting a membrane and optimizing operational conditions for a particular separation process. Table 1.3 summarizes the commonly used transport models of NF membranes, together with the model parameters that must be determined for the models to be applied. A comparison of these models will be presented in the following subsections. Generally, physical sieving is the dominant rejection mechanism in the NF membrane for colloids and large molecules, whereas

TABLE 1.3

Summary of Commonly Used Transport Models for NF Membranes

Model	Transport Mechanism	Model Parameters[a] ($1 < i < n_{solute}$ and $1 < j < n_{solvent}$)
Kedem–Katchalsky model	Diffusion and convection	P_i, L_p, σ_i
Spiegler–Kedem model	Diffusion and convection	P_i, L_p, σ_i
Hagen–Poiseuille model	Convection	r_p, l, ϵ, τ
SHP model	Convection	P_i, L_p, r_p, S_p
TMS model	Electrostatic interactions	P_i, σ_i, ξ
DSPM	Diffusion, convection, and electrostatic interactions	r_p, l, ϵ, τ, ψ

[a] Definition of each model parameter can be found in the following subsections.

the chemistries of solute and membrane become increasingly important for ions and lower molecular weight organics. The mechanisms, however, are still not fully understood.

1.3.1 Irreversible Thermodynamics

Irreversible thermodynamic models describe transport as an irreversible process that continuously produces entropy and dissipates free energy. This class of models is useful, especially when the structure of a membrane is not known and the mechanism of transport within the membrane is not fully understood. Although less information is required to setup the models, one can only expect to obtain little information from them about the membrane transport mechanism.

The first irreversible thermodynamic-based membrane model was developed by Kedem and Katchalsky (1958, 1963). According to the Kedem–Katchalsky model, the solvent flux, J_j and solute flux, J_i of aqueous solutions containing a single solute are expressed by the following equations:

$$J_j = L_p(\Delta p - \sigma_i \Delta \pi) \tag{1.1}$$

$$J_i = P_i \Delta c_i + (1 - \sigma_i) J_j \bar{c}_i \tag{1.2}$$

where L_p is the solvent permeability coefficient, Δp is the permeation pressure, σ_i is the reflection coefficient, corresponding to the solute fraction rejected by the membrane, $\Delta\pi$ is the osmotic pressure difference of water across the membrane, P_i is the solute permeability coefficient, and $\bar{c}_i$ is the average solute concentration in the membrane. The $\Delta\pi$ in Equation 1.1 can be further defined by the van't Hoff equation as follows:

$$\Delta \pi = vRT(c_{i,0} - c_{i,l}) \tag{1.3}$$

where v is the number of ions, R is the gas constant, T is the temperature, and $c_{i,0}$ and $c_{i,l}$ are the solute concentration in the membrane at feed and permeate interface, respectively.

In Equation 1.2, the transport of solutes due to diffusion and convection is represented by the first and second term of the right-hand side of the equation. The solute diffusion is dependent on its concentration while the convection is proportional to the applied pressure. Based on the simple phenomenological transport model, it should be realized that retention is dependent not only on the flux but also on the solute concentration. Spiegler and Kedem (1966) further expressed the flux of solute, J_i in a differential form when a high concentration difference between retentate and permeate exists:

$$J_i = -P'\left(\frac{dc_i}{dx}\right) + (1-\sigma_i)J_j\bar{c}_i \tag{1.4}$$

where P' is the local solute permeability, defined as $P' = P_i\Delta x$. The permeability coefficient of solute, P_i and the reflection coefficient, σ_i can be obtained by fitting of the experimental solute rejection (observed rejection) versus flux, according to the following equation:

$$R = \frac{\sigma_i(1-F)}{1-\sigma_i F} \tag{1.5}$$

where

$$F = \exp\left(-\frac{1-\sigma_i}{P_i}J_j\right) \tag{1.6}$$

One can see from Equation 1.5 that σ_i corresponds to the maximum rejection at infinite volume flux. It can be determined directly from the experimental data of rejection, R as a function of solvent flux, J_i by a best-fit method. The major advantage of these two phenomenological models is that they are simple to apply and have been found to describe the rejection behavior of a variety of solutes, although sometimes it is quite difficult to use them for both predictive and NF membrane characterization purposes (Bitter, 1991).

1.3.2 Steric Hindrance Pore Model and Hagen–Poiseuille Model

The steric hindrance pore (SHP) model is adopted in order to provide a description of the separation properties of membranes with respect to two key parameters, pore radius, r_p and ratio of porosity to thickness of a membrane, A_k/l. Through modifying the pore model, Nakao and Kimura (1982) proposed that the SHP model can be used to calculate r_p and A_k/l if the system only consists of single neutral solute. The parameters can be calculated by equations as follows:

$$\sigma_i = 1 - H_F S_F \tag{1.7}$$

$$P_i = H_D S_D D_i \left(\frac{A_k}{l} \right) \tag{1.8}$$

where H_F and H_D are the steric parameters related to the wall correction factors in the convection coefficient and diffusion coefficient, respectively and S_F and S_D are the distribution coefficients of solute in the convection condition and diffusion condition, respectively. The nomenclatures in Equations 1.7 and 1.8 can be further defined by the following expressions:

$$H_F = 1 + \frac{16}{9}\lambda^2 \tag{1.9}$$

$$H_D = 1 \tag{1.10}$$

$$S_F = (1-\lambda)^2[2-(1-\lambda)^2] \tag{1.11}$$

$$S_D = (1-\lambda)^2 \tag{1.12}$$

where

$$\lambda = \frac{r_i}{r_p} \tag{1.13}$$

where r_i and r_p are the solute and membrane pore radius, respectively. To determine the solute diffusivity where the solute is dilute in the solvent, the Wilke–Chang correlation is often used (Wilke and Chang, 1955).

$$D_i = 1.173 \times 10^{-16} (\varphi_j M_j)^{1/2} \frac{T}{\mu_j V_i^{0.6}} \tag{1.14}$$

where V_i is the solute molar volume at boiling point, T is the temperature, φ_j, M_j, and μ_j are an "association parameter," molecular weight, and viscosity of the solvent, respectively. Several assumptions are made in this pore model in order to determine the membrane properties (Nakao and Kimura, 1982; Lau and Ismail, 2009). They are (1) pore-wall effects on solute moving through the pore are eliminated and (2) there is no discrepancy between the SHP model and the fiction model.

On the other hand, pure solvent flux through uniform cylindrical pores, where no significant concentration gradient is present across the membrane, can be described by the Hagen–Poiseuille equation as follows:

$$J_j = \frac{\varepsilon r_p^2 \Delta P}{8\mu\tau l} \tag{1.15}$$

According to this model, the viscosity (μ) is the only solvent parameter influencing permeation. The membrane properties are represented by the porosity (ϵ), pore radius (r_p), membrane thickness (l), and tortuosity factor (τ). Although this equation is usually used for porous membranes, it is also valid for NF membranes (Marchetti et al., 2014). The Hagen–Poiseuille equation clearly shows the effect of membrane structure on transport and how membrane performance can be improved by varying some specific parameters. It gives a good description of the transport through membranes consisting of circular pores of the same radius, although this is generally not the case in practice.

1.3.3 Teorell–Meyer–Sievers Model

Teorell (1935) and Meyer and Sievers (1936a,b) discussed the membrane phenomenon in an aqueous electrolyte solution and revealed the fundamental mechanism of characteristics such as membrane potential, diffusion coefficient, transport number, electric conductivity, etc. The Teorell–Meyer–Sievers (TMS) model was developed based on the Donnan equilibrium theory and the extended Nernst–Planck equation. This model is a rigorous approach that has been widely used to describe the membrane electrical properties (in the case of a negatively charged membrane) by assuming a uniform radial distribution of fixed charges and mobile species. For 1-1 type of electrolyte (e.g., NaCl), the membrane parameters can be written as follows:

$$\sigma_i = 1 - \frac{2}{(2\alpha - 1)\xi + (\xi^2 + 4)^{0.5}} \tag{1.16}$$

$$P_i = D_i(1 - \sigma_i)(A_k / l) \tag{1.17}$$

where α is the transport numbers of cation in free solution and ξ is the electrostatic parameter. Both nomenclatures can be further defined by the following expressions:

$$\alpha = \frac{D_{cation}}{D_{cation} + D_{anion}} \tag{1.18}$$

$$\xi = \frac{X}{c_i} \tag{1.19}$$

where D_{cation} and D_{anion} are the diffusivity of cation and anion, respectively and X and c_i are the fixed charge density and the electrolyte concentration of feed solution, respectively. However, the fixed charge density of most of the practical charged membranes varies with the electrolyte concentrations, thus the effective fixed charge density ϕX was proposed by Wang et al. (1995a) to replace X as follows:

$$\phi X = \frac{Ac_i^{0.5}}{1 + Bc_i^{0.5}} \tag{1.20}$$

where A and B are the empirical coefficients obtained from the permeation experiments of a salt (Wang et al., 1995b).

1.3.4 Donnan Steric Pore-Flow Model

Of the various models developed to describe the NF separation, the use of the extended Nernst–Planck equation is best known in predicting ion transport (Dresner, 1972). Such an equation describes the transport of ions across the membrane in terms of diffusion and migration, as a result of concentration and electrical potential gradients as well as convection due to the pressure difference across the membrane.

Bowen and Mukhtar in 1996 developed the so-called Donnan steric pore-flow model (DSPM) that is based on the extended Nernst–Planck equation by taking into account the sieving effect.

$$J_i = -K_{i,d} D_i \frac{dc_i}{dx} + K_{i,c} c_{i,o} J_j - \frac{z_i c_{i,o} K_{i,d} D_i}{RT} F \frac{d\Psi}{dx} \tag{1.21}$$

where D_i, $c_{i,0}$, J_i, R, and T were stated in previous equations, z_i is the ion valence, F is the Faraday constant, $d\Psi/dx$ is the electric potential gradient across the membrane, and $K_{i,d}$ and $K_{i,c}$ are the hindrance factors for convention and diffusion, respectively. According to this equation, solute flux is affected by diffusion (driven by a concentration gradient), convection (with the total volume flux), and electromigration (Donnan exclusion). For neutral solute application, the last term on the right side of Equation 1.21 is negligible.

For the mathematical derivation of the DSPM, the following assumptions are usually made (Labbez et al., 2002):

1. The membrane consists of a bundle of identical straight cylindrical pores of radius, r_p and length, l
2. The effective of membrane charge density X is constant throughout the membrane
3. The size of ions and nonionic solutes is expressed as a Stokes radius, r_i

4. The ion fluxes, J_i, ion concentration, $c_{i,0}$, electric potential, Ψ, and permeate volume flux, J_j are all defined in the terms of radially averaged quantities

Rearranging Equation 1.21 will give two differential equations for concentration and electric potential gradients as expressed in Equations 1.22 and 1.23, respectively,

$$\frac{dc_i}{dx} = \frac{J_j}{K_{i,d}D_i}(K_{i,c}c_{i,o} - c_{i,p}) - \frac{z_i c_{i,o}}{RT}F\frac{d\Psi}{dx} \tag{1.22}$$

$$\frac{d\Psi}{dx} = \frac{\sum_{i=1}^{n} \frac{z_i J_j}{K_{i,d}D_i}(K_{i,c}c_{i,o} - c_{i,p})}{\frac{F}{RT}\sum_{i=1}^{n}(z_i^2 c_{i,0})} \tag{1.23}$$

For ions, the application of the DSPM requires an iterative approach to solve these two differential equations describing the concentration and electric potential gradients across the membrane. This will allow the determination of solute retention for the given membrane and feed solution. Calculations can also be performed to determine membrane parameters, provided the feed solution parameters and component retentions as a function of permeate volume flux are known.

References

BCC Research. 2007. Nanotechnology. Report Code: NAN045A. BCC Research, Massachusetts, USA.

BCC Research. 2014. Global Markets and Technology for Nanofiltration. Report Code: NAN045B. BCC Research, Massachusetts, USA.

Bitter, J.G.A. 1991. *Transport Mechanism in Membrane Separation Processes.* Plenum Press, New York, USA.

Bowen, W.R., Mukhtar, H. 1996. Characterisation and prediction of separation performance of nanofiltration membranes. *Journal of Membrane Science.* 112, 263–274.

Cohen, H., Loeb, S. 1977. Industrial waste treatment by means of RO membranes. In: *Reverse Osmosis and Synthetic Membranes*, S. Sourirajan (Ed.), National Research Council Canada, Ottawa, Canada, Chapter 25, pp. 511–525.

Dresner, L. 1972. Some remarks on the integration of the extended Nernst–Planck equations in the hyperfiltration of multicomponent solutions. *Desalination.* 10, 27–46.

Filtration+Separation. 2005. Nanofiltration: Properties and uses. *Filtration+Separation.* September, 16–21.

Hanft, S. 2010. Seawater and Brackish Water Desalination. Market Report: MST052B. BCC Research, Massachusetts, USA.
Kedem, O., Katchalsky, A. 1958. Thermodynamic analysis of the permeability of biological membranes to non-electrolytes. *Biochimica et Biophysica Acta*. 27, 229–246.
Kedem, O., Katchalsky, A. 1963. Permeability of composite membranes. *Transactions of the Faraday Society*. 59, 1918–1953.
Labbez, C., Fievet, P., Szymczyk, A., Vidonne, A., Pagetti, F.J. 2002. Analysis of the salt retention of a titania membrane using the "DSPM" model: Effect of pH, salt concentration and nature. *Journal of Membrane Science*. 208, 315–329.
Lau, W.J., Ismail, A.F. 2009. Theoretical studies on the morphological and electrical properties of blended PES/SPEEK nanofiltration membranes using different sulfonation degree of SPEEK. *Journal of Membrane Science*. 334, 30–42.
Lau, W.J., Ismail, A.F., Misdan, N., Kassim, M.A. 2012. A recent progress in thin film composite membrane: A review. *Desalination*. 287, 190–199.
Li, N.N., Fane, A.G., Winston Ho, W.S., Matsuura, T. 2008. *Advanced Membrane Technology and Applications*. John Wiley & Sons, New Jersey, USA.
Loeb, S., Sourirajan, S. 1964. High flow semipermeable membrane for separation of water from saline solution. US Patent: 3,122,132. May 12.
Marchetti, P., Jimenez Solomon, M.F., Szekely, G., Livingston, A.G. 2014. Molecular separation with organic solvent nanofiltration: A critical review. *Chemical Reviews*. 114, 10735–10806.
Meyer, K.H., Sievers, J.F. 1936b. La permeabilite des membranes I, Theorie de la permeabilite ionique [Membrane permeability. I. Theory of ion permeability]. *Helvetica Chimica Acta*. 19, 649–664.
Meyer, K.H., Sievers, J.F. 1936a. La permeabilite des membranes II, Essis avec des membranes selectives artifielles [Membrane permeability. II. Experiments with artificial selective membranes]. *Helvetica Chimica Acta*. 19, 665–677.
Nakao, S.I., Kimura, S. 1982. Models of membrane transport phenomena and their applications for ultrafiltration data. *Journal of Chemical Engineering of Japan*. 15(3), 200–205.
Paulson, D. 2015. Nanofiltration: The up-and-coming membrane process. *Water Online*. May 18, http://www.wateronline.com.
Schafer, A.I., Fane, A.G., Waite, T.D. 2005. *Nanofiltration: Principles and Applications*. Elsevier, Oxford, UK.
Spiegler, K.S., Kedem, O. 1966. Thermodynamics of hyperfiltration (reverse osmosis): Criteria for efficient membranes. *Desalination*. 1, 311–326.
Strathmann, H. 2011. *Introduction to Membrane Science and Technology*. Wiley-VCH, Weinheim, Germany.
Teorell, T. 1935. An attempt to formulate a quantitative theory of membrane permeability. *Proceedings of the Society for Experimental Biology and Medicine*. 33, 282–385.
Wang, X.-L., Tsuru, T., Togoh, M., Nakao, S., Kimura, S. 1995a. Transport of organic electrolytes with electrostatic and steric-hindrance effects through nanofiltration membranes. *Journal of Chemical Engineering of Japan*. 28, 372–380.
Wang, X.-L., Tsuru, T., Togoh, M., Nakao, S., Kimura, S. 1995b. Evaluation of pore structure and electrical properties of nanofiltration membranes. *Journal of Chemical Engineering of Japan*. 28, 186–192.
Wilke, C.R., Chang, P. 1955. Correlation of diffusion coefficients in dilute solutions. *American Institute of Chemical Engineers Journal*. 1, 264–270.

2

Synthesis of Nanofiltration Membrane

2.1 Overview of NF Membrane Synthesis

In this chapter, three main methods to synthesize polymeric NF membranes will be described. They are the IP technique that is used to prepare PA thin film (nano)composite membrane, single-step phase inversion method for asymmetric NF membrane making, and layer-by-layer (LBL) assembly method for multilayer NF membrane fabrication.

2.2 PA Thin Film (Nano)Composite Membrane via IP Technique

The concept of IP was first introduced by Morgan in the 1960s (Morgan, 1965). It is generally agreed that the contribution of the IP technique to the development of membrane science and technology is extremely significant, making it almost equivalent to the historical announcement of Loeb–Sourirajan's asymmetric membrane made of the phase inversion technique. By employing the IP technique, an ultrathin selective PA layer with several hundreds of nanometers in thickness is able to form over the surface of the microporous support membrane, producing a TFC membrane with a good combination of water flux and solute rejection. PA is polymer where the monomer units are connected together by the amide group (–C(=O)–NH–). It can be easily synthesized by IP between polyfunctional amine monomer and polyacyl chloride monomer at the interface between two immiscible solvents with hydrochloric acid (HCl) created as a byproduct.

Conventionally, the interfacially polymerized PA layer is established by immersing the support membrane in an aqueous solution containing amine monomer (with concentration mostly between 1 and 3 w/v%) followed by organic solution of acyl chloride monomer (mostly between 0.05 and 0.2 w/v%). The membrane is then subject to a heat treatment process at temperature in the range of 70–90°C to densitify the polymerization properties of the PA layer.

Figure 2.1 shows the steps involved in preparing a small size of TFC membrane for lab-scale evaluation. In brief, a microporous membrane (generally made of polysulfone [PSF] or polyethersulfone [PES]) is immersed in pure water overnight (Figure 2.1a). A rubber gasket and a plastic frame are used and placed on top of the support membrane surface in which clips are used to hold the plate–membrane–gasket–frame stack together (Figure 2.1b). To start the IP process, an aqueous solution with a certain amount of amine monomer (prepared in the unit of w/v%, i.e., g amine monomer/100 mL

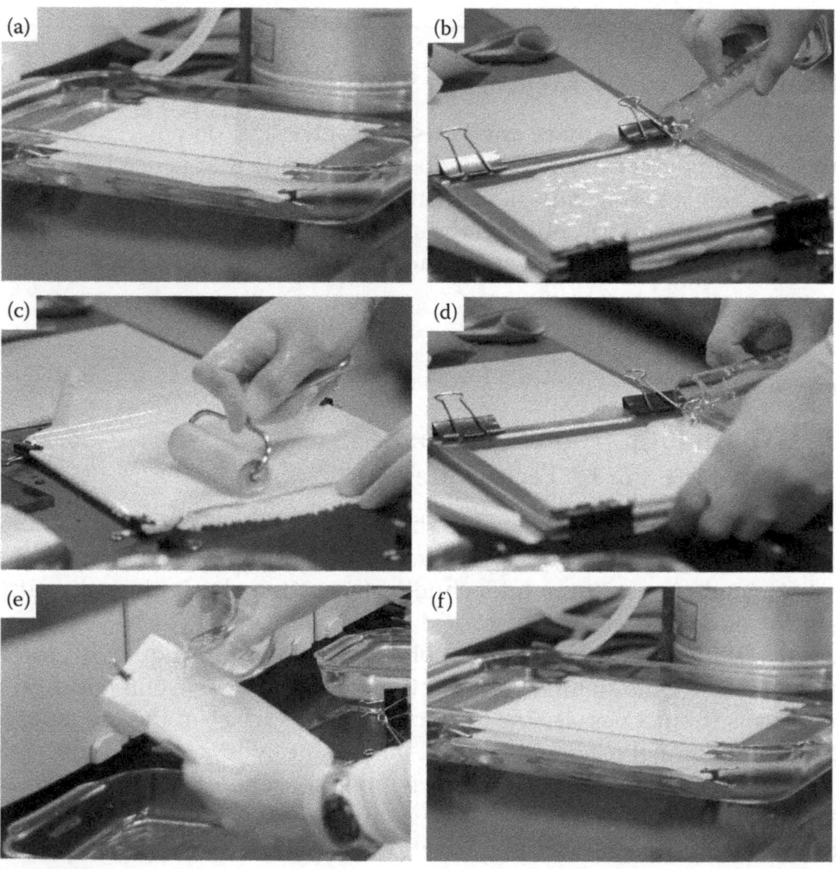

FIGURE 2.1
Steps in the protocol used to prepare PA TFC membranes, (a) immerse a microporous support membrane in pure water overnight; (b) contact the membrane surface with an aqueous amine solution for several minutes; (c) remove aqueous droplets from the membrane surface with a soft rubber roller; (d) contact the membrane surface with an organic solution of acyl chloride; (e) rinse the membrane surface with organic solvent; and (f) store the TFC membrane in pure water. (Reprinted from *Journal of Membrane Science*, 403–404, Xie, W. et al., Polyamide interfacial composite membranes prepared from m-phenylene diamine, trimesoyl chloride and a new disulfonated diamine, 152–161, Copyright 2012, with permission from Elsevier.)

pure water) is poured into the frame (Figure 2.1b) and allowed to contact the membrane surface for several minutes before draining the excess aqueous solution. The frame and gasket are disassembled and a soft rubber roller is used to remove the residual solution (Figure 2.1c). This step is necessary as the existence of any aqueous droplets could later form defects on the PA layer. Other than a rubber roller, an air knife or tissue paper sometimes is used to remove the excess solution. A PA layer will start to develop as soon as the organic solution (usually *n*-hexane or cyclohexane) that contains the secondary monomer is poured on top of the support membrane (Figure 2.1d). Once the interaction between two active monomers is completed, the organic solution is drained from the frame, and the frame and gasket are dissembled again. The TFC membrane surface is usually rinsed using the same organic solution used in Figure 2.1d to wash away residual monomers (Figure 2.1e). After the membrane is dried in air at ambient conditions, it is transferred to a pure water container (Figure 2.1f). In most cases, a heat treatment is performed by placing the TFC membrane in an oven before storing it in a water container. This step is required in order to ensure the complete establishment of a PA thin layer on the surface of the support membrane. With respect to the residence time of the aqueous and organic solution on the support membrane surface, there is no definite answer as the interactions between monomers and membrane surface are governed by many factors. However, the residence time of the aqueous solution is generally longer than the organic solution to allow more amine monomers to penetrate into the pores of the support membrane for a higher degree of PA cross-linking (CL).

There are some cases where the sequence of immersion process is reversed (i.e., the organic solution is introduced before the aqueous solution) for the support membrane that is made of polymer with a relatively higher hydrophobic such as polyvinylidene fluoride (Lu et al., 2002). This reverse process is used because the hydrophobic substrate tends to interact better with the organic solution of acyl chloride monomer in comparison to the aqueous amine solution. Wu et al. (2010) on the other hand applied the organic solution before the aqueous solution in the fabrication process of a thin film nanocomposite (TFN) membrane due to the poor dispersion of multi-walled carbon nanotube (MWCNT) in the nonpolar solvent of the organic phase. To ensure good dispersion of the MWCNT in solution and its even distribution in the PA layer, the MWCNT was introduced in the aqueous solution and used as secondary solution during the IP process. In addition to the two conventional immiscible solutions, the use of tertiary amine solution has been previously reported to further react with unreacted acid chloride groups on the surface of the PA layer (Zou et al., 2010). The modified approach was reported to yield a membrane surface with a larger amount of amino groups ($-NH_2$), causing the membrane to have better antifouling properties than that of the conventionally prepared membrane.

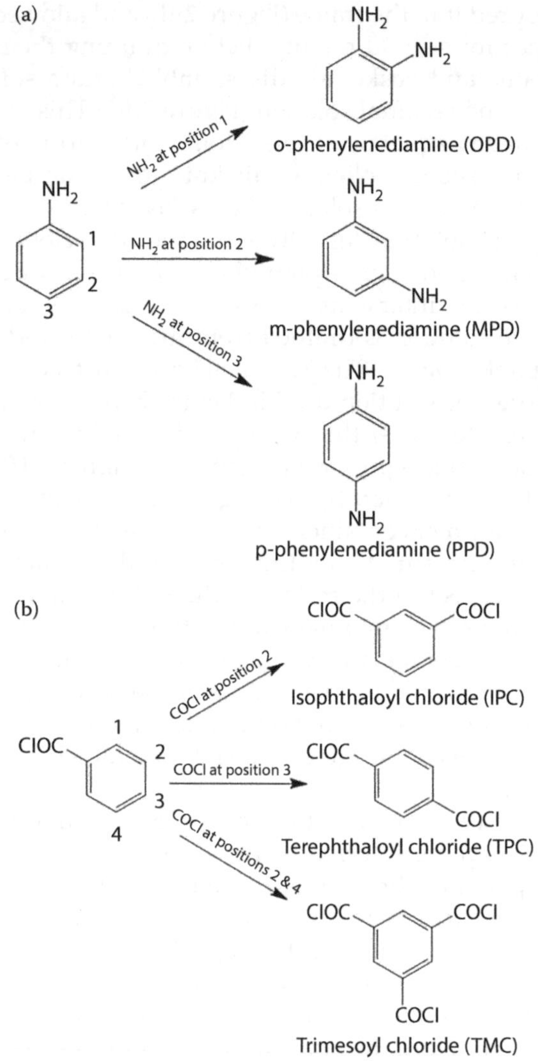

(a)

(b)

FIGURE 2.2
Organic structures of (a) aromatic diamine monomer and (b) bi-/tri-acyl chloride monomer used in synthesizing PA.

The characteristics of PA are varied depending on the type of amine and acyl chloride monomers selected during the IP process. Figure 2.2 shows the most frequently used aromatic diamine monomer and bi-/tri-acyl chloride monomers in preparing the cross-linked aromatic PA layer. The degree of PA CL depends not only on the organic structure of the monomers used but also on factors such as the concentration of active reactant in the aqueous/organic solution, type of hydrocarbon solvent, reaction time, presence of additive/surface, etc. The possible PA structures formed from the reaction

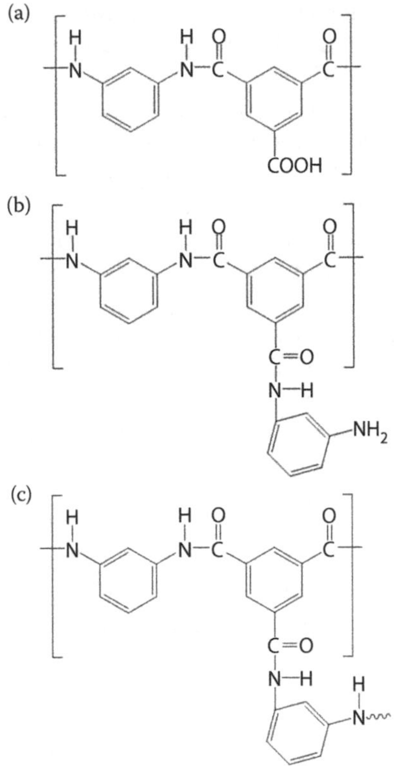

FIGURE 2.3
Possible chemical structures of PA prepared from MPD and TMC, (a) linear structure with a pendant –COOH group from the hydrolysis of –COCl, (b) cross-linked with a pendant $–NH_2$ group, and (c) totally cross-linked polymer chain.

of *m*-phenyldiamine (MPD) with TMC are shown in Figure 2.3. It is generally agreed that the poly(MPD–TMC) chain is to contain both the cross-linked structure and non-cross-linked (i.e., linear) structure. The carboxylic acid functional group (–COOH) arises due to the partial hydrolysis of the acyl chloride unit of TMC during the IP process. In order to grasp the CL reaction rate of the PA formed, characterizations using Fourier transform infrared (FTIR) spectrometer and X-ray photoelectron spectroscope (XPS) are generally performed.

Similar to the TFC membrane, the TFN membrane that incorporates nanomaterials within the PA dense layer can also be produced using the IP technique. Research interest in TFN membranes has subsequently increased since the first reports from Hoek's research group (Jeong et al., 2005, 2007). The performance of the TFN membrane, however, varies depending on the features of nanomaterials used. These include particle size, hydrophilicity, charge properties, and pore channels. It is often reported that the properties of the interfacially polymerized layer can be improved by introducing nanomaterial into the PA selective layer, provided the amount of nanomaterial added is properly controlled.

To prepare the TFN membrane, the nanomaterials could be introduced into either aqueous or organic solutions. However, most research has preferred to introduce nanomaterials that are in hydrophilic nature into amine aqueous solution. This could be mainly due to the difficulties in producing a homogenous mixture in the nonpolar organic phase. The high surface energy of inorganic nanomaterials coupled with high inter-particle interactions are the main reasons causing them to aggregate easily in the organic phase and form defects in the PA layer. Figure 2.4 illustrates a new approach to graft the PA top surface with silver (Ag) nanoparticles after the IP process (Yin et al., 2013). This approach chemically immobilizes Ag nanoparticles on the TFC membrane surface by reacting the PA layer with NH_2–$(CH_2)_2$–SH ethanol solution (20 mM) followed by incubating with an Ag nanoparticles suspension (0.1 mM) for 12 h. Besides showing enhanced water flux (compared with TFC membrane), the resultant TFN membrane also demonstrated excellent antibacterial ability to inhibit *Escherichia coli* growth. It is claimed by the authors that this grafting approach was very effective at reducing Ag nanoparticle leaching during filtration due to the strong covalent bonding formed between the thiol groups and Ag nanoparticles as confirmed by Raman spectra.

Other than static IP process, the dynamic IP process has been attempted to form a thin PA on top of the microporous substrate (An et al., 2012). The process is performed using a spin coater in which the organic solution is dispensed onto the middle part of the substrate saturated with amine monomers. The solution then spreads away from the axis of rotation toward the substrate outer edge due to the centrifugal force, leading to the formation of thinner PA layer (around 200 nm) than that produced by static process (>400 nm). Even though the dynamic IP technique seems to be better in producing a thinner PA layer, this approach is seldom adopted by scientists in preparing TFC membrane. One of the main reasons is the limitation of the commercial spin coater for the production of a large membrane area in a continuous manner.

2.3 Asymmetric Membrane via Phase Inversion Technique

The phase inversion method, also known as the polymer precipitation process is the process of the transformation of polymer solution from a liquid state into a solid state (i.e., membrane formed). The transformation of phase state can be achieved in several ways such as immersion in a bath of nonsolvent water (immersion precipitation), casting hot polymer solution on a cool cast film (thermal gelation), contacting the polymer solution with a

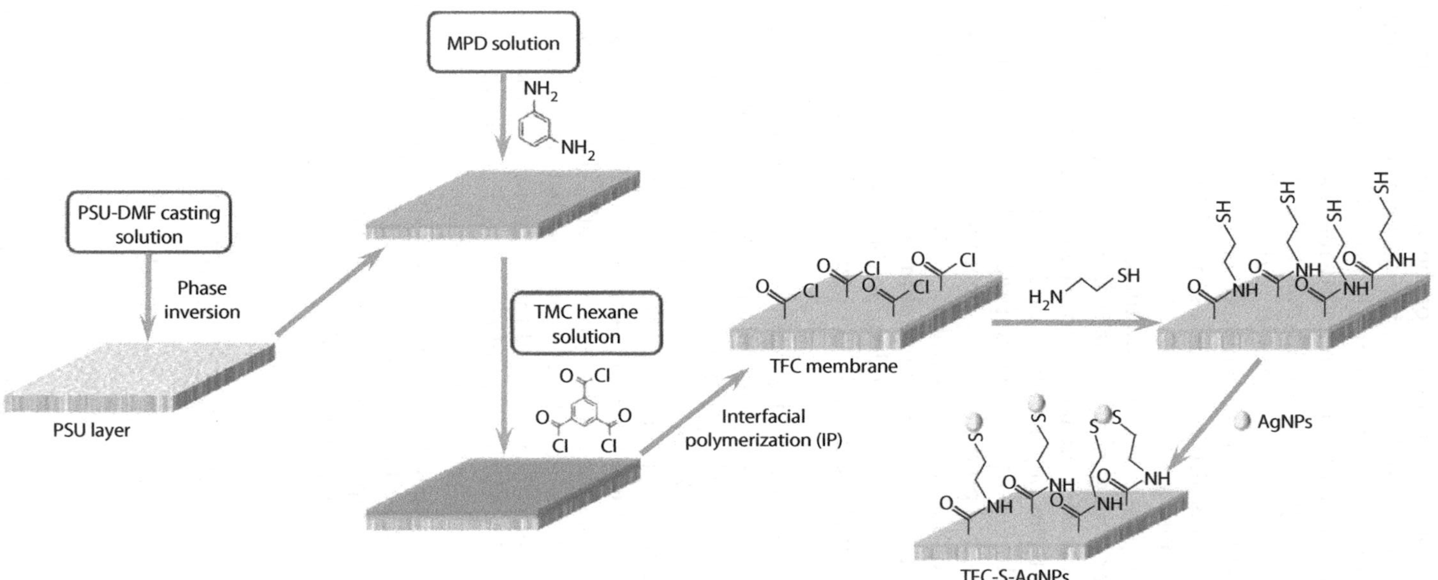

FIGURE 2.4
Schematic diagram of immobilization of Ag nanoparticles onto the top PA surface of TFC membrane. (Reprinted from *Journal of Membrane Science*, 441, Yin, J. et al., Attachment of silver nanoparticles (AgNPs) onto thin-film composite (TFC) membranes through covalent bonding to reduce membrane biofouling, 73–82, Copyright 2013, with permission from Elsevier.)

vapor of non-solvent, or evaporating one of the solvents used in a polymer solution (solvent evaporation). By taking into account the simplicity of the procedure and cost of fabrication, immersion precipitation using water as non-solvent is the most commonly practiced method. Asymmetric membrane prepared via a dry–wet phase inversion method has a wide range of applications in the fields of separation and purification, ranging from aqueous-based media to organic solvent and gas separation processes. This phase inversion method is also generally used to fabricate the microporous substrate for TFC membrane.

Using the dry–wet phase inversion method, asymmetric NF membrane in either flat sheet or hollow fiber format can be produced. The formation of flat sheet and hollow fiber NF membrane seems to be similar, but, in fact, they are different in terms of penetration of the non-solvent from the "free" side of membrane. The conditions used to prepare flat sheet membrane cannot be simply extended to produce hollow fiber membrane. It is well known that the preparation of hollow fiber membrane involves a greater number of factors compared to that of flat sheet membrane during membrane formation. Its properties are strongly dependent on the large number of spinning parameters that govern phase inversion kinetics and interfacial mass transfer during the spinning process.

Unlike flat sheet membrane making where only one coagulation process is involved, hollow fiber membrane fabrication involves two coagulation processes at the same time. One is through the continuous flow of bore fluid in the lumen of the fiber, while the other is through the outside contact with the coagulant as the hollow fiber enters the coagulation bath. In flat sheet membrane preparation, as illustrated in Figure 2.5, there is only one interface between the polymer solution and non-solvent bath. The penetration of the non-solvent occurs from the side of the membrane that is not attached to the support (glass plate). Since the volume of the non-solvent in the coagulation bath is large compared to that of the polymer film, the solvent/non-solvent exchange is very fast and the membrane is formed instantaneously after immersion. On the other hand, there are situations where the penetration of the non-solvent could occur from both sides of the flat film if nonwoven polyester is used as the supporting layer. In this case, there are two interfaces between the solution and non-solvent bath.

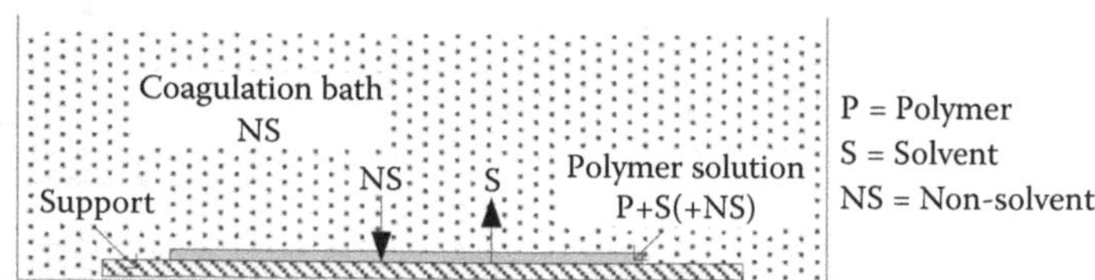

FIGURE 2.5
Schematic representation for preparation of flat sheet NF membrane by immersion precipitation.

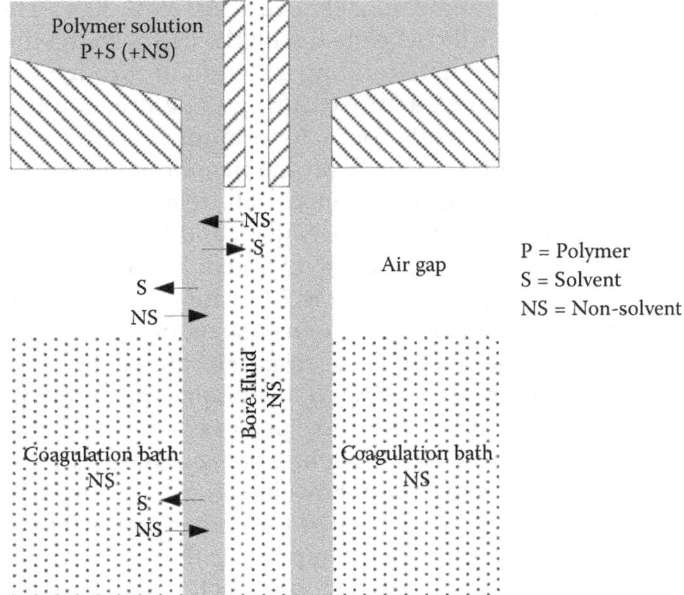

FIGURE 2.6
Schematic diagram of dry-jet wet spinning process for hollow fiber by immersion precipitation.

Compared to the asymmetric flat sheet membrane, the hollow fiber membrane formation by immersion precipitation is much more complicated. In most spinning processes (using the dry-jet wet spinning technique), there are three stages of diffusion induced phase separation as illustrated in Figure 2.6. These include (a) vapor penetration of non-solvent at the outer surface in the air gap, (b) immersion precipitation at the outer surface, once the nascent membrane has passed the air gap and enters the coagulation bath, and (c) immersion precipitation from the lumen side of fiber through diffusional exchange with a bore liquid.

All of these phase preparation conditions have been well recognized as the main factors that determine the morphology of the membrane, in addition to the dope formulation used. Research works have been intensively carried out by membrane scientists over the years to investigate the influences of various spinning conditions on the rate of diffusion between the solvent and non-solvent in order to produce the desired membrane performance that is best suited to particular membrane applications. Such factors affecting the characteristics, morphology, and properties of the resultant hollow fiber membrane can be found easily in the literature and will not be discussed in detail here.

Over the years, there have been a number of studies reported on the fabrication of asymmetric NF membranes through a single-step phase inversion process (Bowen et al., 2001, 2002; Ismail and Hassan, 2004, 2007;

Wang and Chung, 2006; Wang et al., 2007; Lau and Ismail, 2009). These membranes have been successfully developed with the use of charged polymeric materials as the main membrane-forming materials or through the addition of a small amount of charged additive during dope preparation. The dissociation of chemical functional groups, for example, sulfonic acid ($-SO_3H$) and carboxylic acid group (–COOH), on the membrane top surface is the reason why these membranes can demonstrate relatively high separation efficiency against divalent salts such as Na_2SO_4 and $MgSO_2$.

Wang and Chung (2006) developed charged hollow fiber NF membranes directly from polybenzimidazole (PBI) for chromate removal due to the self-charging ability of PBI in aqueous solution. PBI becomes self-charged because the adjacent benzene ring delocalizes the proton of the imidazole group. In comparison to the IP technique, the preparation of NF membranes via a single-step fabrication process is relatively simple and involves less preparation parameters while being able to exhibit the positive features of the pure component. Bowen et al. (2001) have reported the effect of a small percentage of sulfonated poly(ether ether) ketone (SPEEK) as a secondary polymeric additive for improving the performances of PSF-based NF membranes. The increases in salt rejection and water flux of the blend membrane were attributed to the increased membrane charge density and hydrophilicity upon SPEEK addition. Lau and Ismail (2009) on the other hand studied the potential of hollow fiber PES membrane incorporated with SPEEK and found that SPEEK was miscible with PES polymer in *n*-methyl-2-pyrrolidone (NMP) solvent. Excessive use of SPEEK, however, tended to create defects on the membrane morphology which led to lower salt separation efficiency.

Table 2.1 presents the properties of several asymmetric NF membranes made via the phase inversion process using dope solutions containing either polymeric or inorganic additives. As can be seen, there is a significant variation in the water flux and salt rejection of the asymmetric membranes synthesized. The membrane pure water permeabilities in particular are found to be several orders lower than those of the famous TFC membranes when both membranes are evaluated under the same conditions. Nevertheless, the unique advantage of asymmetric membranes is their excellent tolerance to chlorine. Chlorine is generally known as the main component attacking the PA structure of TFC membrane, forming N-chlorinated amide in the initial step. The process is followed by a nonreversible reaction, that is, ring-chlorination through intramolecular rearrangement of the chlorine atom into the aromatic ring of the diamine moiety via Orton rearrangement.

Although asymmetric NF membrane prepared via the phase inversion method has been studied for quite a long time, several major drawbacks of this kind of membrane structure have kept it less competitive in the market. One of the major drawbacks is its low water permeability that resulted from the thick separating layer. This, as a consequence, requires higher operating pressure and cost to achieve a similar water flux of the TFC membrane.

TABLE 2.1
Properties of Asymmetric NF Membranes Prepared via Phase Inversion Method Using Water as Non-Solvent

Dope Formulation[a]					
Main Polymer (wt%)	Polymeric/Inorganic Additive (wt%)	Configuration[b]	Testing Conditions	Optimized Membrane Properties[c]	References
PES (20%)	SPEEK with 77% degree of sulfonation (2% or 4%)	HF	Filtr. mode: Cross-flow Pressure: 6 bar Temp.: Ambient Salt conc.: 1000 ppm	PWF: ~2.05 L m^{-2} h^{-1} Pore size: 1.02 nm Na_2SO_4 rej.: ~92% NaCl rej.: ~48%	Lau and Ismail (2009)
PES (20.34%)	PVPK30 (1%–9%)	FS	Filtr. mode: Dead-end Pressure: 7 bar Temp.: Ambient Salt conc.: 500 ppm	PWF: ~23.4 L m^{-2} h^{-1} Pore size: 0.55 nm NaCl rej.: ~36%	Ismail and Hassan (2007)
PBI (21.6%)	LiCl (1.7%)	HF	Filtr. mode: Cross-flow Pressure: 5–20 bar Temp.: Ambient Salt conc.: 3.4 mol m^{-3}	PWF: ~1.86 L m^{-2} h^{-1} bar^{-1} Pore size: 0.348 nm Na_2SO_4 rej.: ~77% (at 15 bar) NaCl rej.: ~38% (at 15 bar)	Wang and Chung (2006)
PSF (15%)	AgNPs (0.5%–2%)	FS	Filtr. mode: Dead-end Pressure: 12 bar Temp.: 23°C	PWF: ~135 L m^{-2} h^{-1} Pore size: N/A Salt rej.: N/A	Andrade et al. (2015)[d]

(Continued)

TABLE 2.1 (*Continued*)

Properties of Asymmetric NF Membranes Prepared via Phase Inversion Method Using Water as Non-Solvent

Dope Formulation[a]					
Main Polymer (wt%)	**Polymeric/Inorganic Additive (wt%)**	**Configuration[b]**	**Testing Conditions**	**Optimized Membrane Properties[c]**	**References**
PPEES (17%)	TiO_2 (0.02%–0.5%)	FS	Filtr. mode: Dead-end Pressure: 6 bar Temp.: N/A Salt conc.: 1000 ppm	WF: 49.95 L m^{-2} h^{-1} Pore size: N/A Na_2SO_4 rej.: ~73%	Mobarakabad et al. (2015)
PES (18%)	CNTs (0.01%–1%)	FS	Filtr. mode: N/A Pressure: 4 bar Temp.: N/A Salt conc.: 200 ppm	WF: 38.9 L m^{-2} h^{-1} Pore size: N/A Na_2SO_4 rej.: 87% NaCl rej.: 38%	Wang et al. (2015)

[a] PES—polyethersulfone, SPEEK—sulfonated polyether ether ketone, PVP—polyvinylpyrrolidone, PBI—polybenzimidazole, LiCl—lithium chloride, PSF—polysulfone, AgNPs—silver nanoparticles, PPEES—poly(1,4-phenylene ether-ether-sulfone), TiO_2—titanium dioxide, CNTs—carbon nanotubes.

[b] HF—hollow fiber, FS—flat sheet.

[c] PWF—pure water flux, WF—water flux.

[d] This paper only reported the antibacterial activity of NF membrane without providing salt rejection capability.

2.4 Multilayer Membrane via Layer-by-Layer Assembly Method

Multilayer NF membrane prepared via the LBL assembly technique can be achieved by alternating electrostatic adsorption of cationic and anionic polyelectrolytes (PEs) on the surface of a porous supporting membrane. Provided that each adsorption step leads to charge inversion of the surface, the subsequent deposition could form a layered complex, a so-called self-assembled PE complex (PEC), stabilized by strong electrostatic forces. The LBL assembly technique represents an easy, environmentally sound method for preparation of polymer films with thickness in the nanometer range.

The multilayer membrane for NF process was first demonstrated by Stanton et al. (2003) by alternating PE deposition on porous alumina support. The resultant NF membrane was reported to achieve promising water flux of 1–2 $m^3\ m^{-2}\ day^{-1}$ (equivalent to 41.7–83.3 $L\ m^{-2}\ h^{-1}$) at 4.8 bar with $MgSO_4$ rejection recorded at 96%. Figure 2.7 illustrates the process of preparing one bilayer on the top of the charged surface by simply exposing the substrate to a solution of an oppositely charged PE followed by water rinsing to remove excess and weakly adsorbed PE. Since the surface charge is reversed after the first adsorption step, the substrate is exposed to a solution of a PE having the same charge as the initial substrate in the next step, again followed by water rinsing. This process can be repeated many times until the target number of bilayers is achieved.

The selection of substrate is mainly based on surface properties such as roughness and charge density. A low surface roughness and a high surface charge density are prerequisites for a good support for LBL assembly. However, the surface charge can be either inherent or created by surface

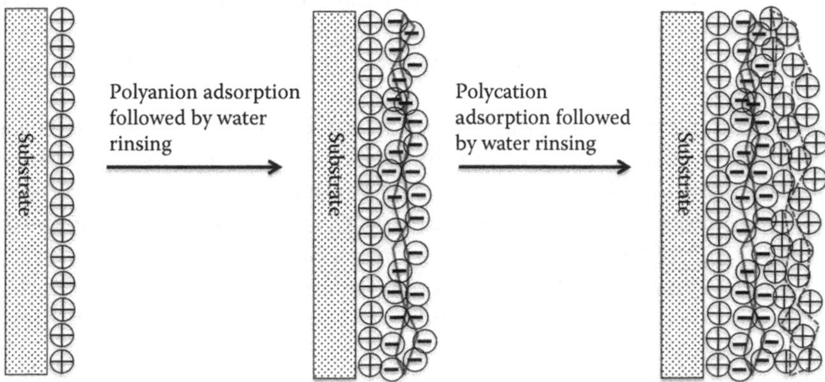

FIGURE 2.7
Illustration of the LBL assembly technique on the charged surface using consecutive adsorption of polyanions and polycations.

treatment. There are several methods available to perform LBL coating. Among them, dip coating is the most commonly used method in preparing multilayer membranes in comparison to spray and spin coating.

As documented in the literature, the polyanion/polycation that have been previously used for multilayer NF membrane preparation are poly(diallyldimethylammonium chloride) (PDADMAC)/poly(styrene sulfonate) (PSS), poly(vinylamine) (PVAm)/poly(vinyl sulfate potassium salt) (PVS), PDADMAC/polyacrylic acid (PAA), sodium carboxymethyl cellulose (CMCNa)/polyethylenimine, poly(allyl amine) hydrochloride (PAH)/PAA, and PSS/PAH. In addition to polycations and polyanions, other (multi)functional macromolecules can be incorporated into the multilayers, leading to enhanced properties of the layers. An example of these macromolecules is polyzwitterions, polymers that contain both positive and negative moieties on the same monomer. The unique properties of polyzwitterions have made them interesting materials for developing an NF membrane with enhanced properties. Figure 2.8 shows the formation of a trilayer that consists of PDADMAC/poly N-(3-sulfopropyl)-N-(methacryloxyethyl)-N,N-dimethylammonium betaine (PSBMA)/PSS on a charged hollow fiber membrane surface. With the incorporation of the polyzwitterion (PSBMA) in the multilayer, the resultant NF membrane could achieve excellent retentions for both positively and negatively charged micropollutants, a behavior that is attributed to dielectric exclusion of the solutes (de Grooth et al., 2014). Furthermore, the NF membrane made of optimum coating conditions exhibited 98% Na_2SO_4 and 42% NaCl rejection with permeabilities recorded at 3.7–4.5 L m^{-2} h^{-1} bar^{-1}. Besides the organic polymers, inorganic nanoparticles can also be deposited on the surface of the substrate via LBL assembly to construct NF membranes. Some of the nanoparticles that have been used are silica and titanium dioxide nanoparticles (Kim et al., 2013; Escobar-Ferrand et al., 2014). For more details on the coating parameters that can influence the multilayer properties such as structure, stability, and performance, one can refer to the comprehensive review articles written by Xu et al. (2015) and Joseph et al. (2014) in recent years.

Instead of the static assembly method as described earlier, the cross-flow dynamic assembly method has also been attempted by placing the substrate in a cell in which the top surface is flushed with dynamic solutions of polyanions and polycations under pressure condition. For the case of PDADMAC/PSS bilayers on a PSF substrate, Su et al. (2012) reported the cross-flow dynamic assembly method could achieve a relatively higher rejection with less number of bilayers compared with that of the static assembly method, but the flux is relatively smaller. Considering the pros and cons of the two kinds of assembly methods, a novel kind assembly method termed as cross-flow dynamic–static assembly was proposed (Su et al., 2012). The procedure of the dynamic–static LBL preparation method is essentially like the dynamic procedure, except that during

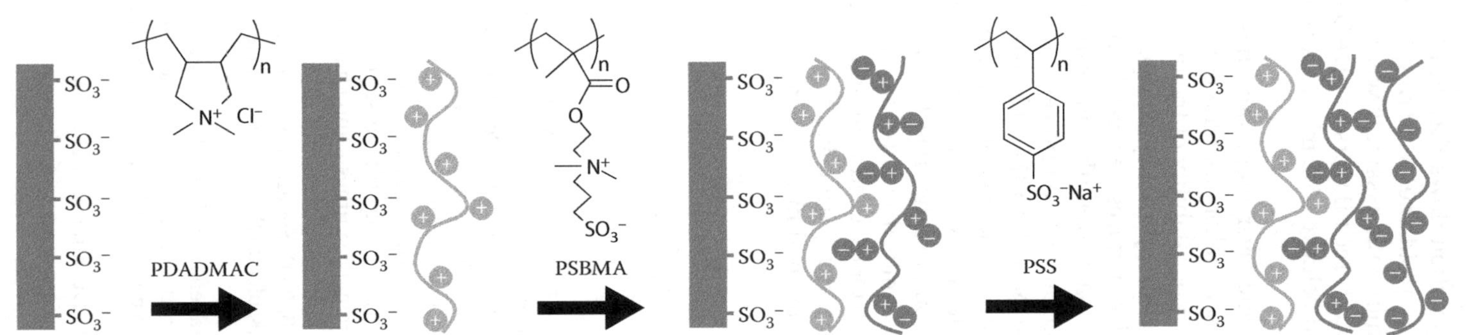

FIGURE 2.8
Schematic representation of the coating of a charged hollow fiber membrane with a trilayer of PDADMAC (+)/PSBMA (z)/PSS (–) on a charged hollow fiber membrane made of sulfonated PES. (Reprinted with permission from de Grooth, J. et al., Charged micropollutant removal with hollow fiber nanofiltration membranes based on oolycation/polyzwitterion/polyanion multilayers. *ACS Applied Materials and Interfaces.* 6, 17009–17017. Copyright 2014 American Chemical Society.)

the assembly of PSS, the PSS solution flows through the membrane surface at atmospheric pressure to avoid the occurring of any permeation. Results showed that the dynamic–static assembly was more effective than the other two methods and with three bilayers of PDADMAC/PSS established, the NF membrane could achieve 55 L m^{-2} h^{-1} water flux and about 90% SO_4^{2-} at 10 bar.

It is demonstrated that the multilayer membrane has potential for NF applications, but it still faces great challenges. One major restriction of the LBL assembled membrane is the poor stability of the bilayer constructed by the electrostatic interaction between oppositely charged PEs. Studies have shown that the LBL assembled structures can easily be destroyed under severe conditions, such as high ionic strength and chlorine treatment that are always involved in the water purification process. Although the glutaraldehyde CL process can be performed on the LBL assembled membrane, it is still found insufficient under some conditions (Ichinose et al., 1999). Furthermore, the membrane water flux tends to decrease significantly with increasing bilayer numbers, although higher rejection is obtained. Xu et al. (2015) further stated that LBL assembled membranes face great challenges in achieving the same level of currently used commercial TFC membranes, that is, good balance between water permeability and solute selectivity. The utilization of this assembly method might be more suitable for surface modifications of TFC membrane.

References

An, Q., Hung, W., Lo, S., Li, Y., De Guzman, M. 2012. Comparison between free volume characteristics of composite membranes fabricated through static and dynamic interfacial polymerization processes. *Macromolecules*. 45, 3428–3435.

Andrade, P.F., de Faria, A.F., Oliveira, S.R., Arruda, M.A.Z., Gonçalves, M.C. 2015. Improved antibacterial activity of nanofiltration polysulfone membranes modified with silver nanoparticles. *Water Research*. 81, 333–342.

Bowen, W.R., Doneva, T.A., Yin, H.B. 2001. Polysulfone-sulfonated poly(ether ether) ketone blend membranes: Systematic synthesis and characterization. *Journal of Membrane Science*. 181, 253–263.

Bowen, W.R., Doneva, T.A., Yin, H.B. 2002. Separation of humic acid from a model surface water with PSU/SPEEK blend UF/NF membranes. *Journal of Membrane Science*. 206, 417–429.

de Grooth, J., Reurink, D.M., Ploegmakers, J., de Vos, W.M., Nijmeijer, K. 2014. Charged micropollutant removal with hollow fiber nanofiltration membranes based on oolycation/polyzwitterion/polyanion multilayers. *ACS Applied Materials and Interfaces*. 6, 17009–17017.

Escobar-Ferrand, L., Li, D., Lee, D., Durning, C.J. 2014. All-nanoparticle layer-by-layer surface modification of micro- and ultrafiltration membranes. *Langmuir*. 30, 5545–5556.

Ichinose, I., Mizuki, S., Ohno, S., Shiraishi, H., Kunitake, T. 1999. Preparation of cross-linked ultrathin films based on layer-by-layer assembly of polymers. *Polymer Journal.* 31, 1065–1070.

Ismail, A.F., Hassan, A.R. 2004. The deduction of fine structural details of asymmetric nanofiltration membranes using theoretical models. *Journal of Membrane Science.* 231, 25–36.

Ismail, A.F., Hassan, A.R. 2007. Effect of additive contents on the performances and structural properties of asymmetric polyethersulfone (PES) nanofiltration membranes. *Separation and Purification Technology.* 55, 98–109.

Jeong, B.-H., Hoek, E.M.V., Yan, Y., Subramani, A., Huang, X., Hurwitz, G., Ghosh, A.K., Jawor, A. 2007. Interfacial polymerization of thin film nanocomposites: A new concept for reverse osmosis membranes. *Journal of Membrane Science.* 294, 1–7.

Jeong, B.-H., Subramani, A., Yan, Y., Hoek, E.M.V. 2005. Antifouling thin film nanocomposite (Tfnc) membranes for desalination and water reclamation. In: *2005 AIChE Annual Meeting and Fall Showcase,* Cincinnati, Ohio, October 30, 2005 to November 4, 2005, Code 6692.

Joseph, N., Ahmadiannamini, P., Hoogenboom, R., Vankelecom, I.F.J. 2014. Layer-by-layer preparation of polyelectrolyte multilayer membranes for separation. *Polymer Chemistry.* 5, 1817–1831.

Kim, J., Sotto, A., Chang, J., Nam, D., Boromand, A., der Bruggen, B.V. 2013. Embedding TiO_2 nanoparticles versus surface coating by layer-by-layer deposition on nanoporous polymeric films. *Microporous and Mesoporous Materials.* 173, 121–128.

Lau, W.J., Ismail, A.F. 2009. Theoretical studies on the morphological and electrical properties of blended PES/SPEEK nanofiltration membranes using different sulfonation degree of SPEEK. *Journal of Membrane Science.* 334, 30–42.

Lu, X., Bian, X., Shi, L. 2002. Preparation and characterization of NF composite membrane. *Journal of Membrane Science.* 210, 3–11.

Mobarakabad, P., Moghadassi, A.R., Hosseini, S.M. 2015. Fabrication and characterization of poly(phenylene ether–ether sulfone) based nanofiltration membranes modified by titanium dioxide nanoparticles for water desalination. *Desalination.* 365, 227–233.

Morgan, P.W. 1965. Condensation polymers: By interfacial and solution methods, *Polymer Reviews,* Vol. 10, Wiley, New York, 19–64.

Stanton, B.W., Harris, J.J., Miller, M.D., Bruening, M.L. 2003. Ultrathin, multilayered polyelectrolyte films as nanofiltration membranes. *Langmuir.* 19, 7038–7042.

Su, B., Wang, T., Wang, Z., Gao, X., Gao, C. 2012. Preparation and performance of dynamic layer-by-layer PDADMAC/PSS nanofiltration membrane. *Journal of Membrane Science.* 423–424, 324–331.

Wang, K.Y., Chung, T.S. 2006. Fabrication of polybenzimidazole (PBI) nanofiltration hollow fiber membranes for removal of chromate. *Journal of Membrane Science.* 281, 307–315.

Wang, K.Y., Chung, T.S., Qin, J.J. 2007. Polybenzimidazole (PBI) nanofiltration hollow fiber membranes applied in forward osmosis process. *Journal of Membrane Science.* 300, 6–12.

Wang, L., Song, X., Wang, T., Wang, S., Wang, Z., Gao, C. 2015. Fabrication and characterization of polyethersulfone/carbon nanotubes (PES/CNTs) based mixed matrix membranes (MMMs) for nanofiltration application. *Applied Surface Science.* 330, 118–125.

Wu, H., Tang, B., Wu, P. 2010. MWNTs/polyester thin film nanocomposite membrane: An approach to overcome the trade-off effect between permeability and selectivity. *The Journal of Physical Chemistry C*. 114, 16395–16400.
Xie, W., Geise, G.M., Freeman, B.D., Lee, H.-S., Byun, G., McGrath, J.E. 2012. Polyamide interfacial composite membranes prepared from m-phenylene diamine, trimesoyl chloride and a new disulfonated diamine. *Journal of Membrane Science*. 403–404, 152–161.
Xu, G.-R., Wang, S.-H., Zhao, H.-L., Wu, S.-B., Xu, J.-M., Li, L., Liu, X.-Y. 2015. Layer-by-layer (LBL) assembly technology as promising strategy for tailoring pressure-driven desalination membranes. *Journal of Membrane Science*. 493, 428–443.
Yin, J., Yang, Y., Hu, Z., Deng, B. 2013. Attachment of silver nanoparticles (AgNPs) onto thin-film composite (TFC) membranes through covalent bonding to reduce membrane biofouling. *Journal of Membrane Science*. 441, 73–82.
Zou, H., Jin, Y., Yang, J., Dai, H., Yu, X., Xu, J. 2010. Synthesis and characterization of thin film composite reverse osmosis membranes via novel interfacial polymerization approach. *Separation and Purification Technology*. 72, 256–262.

3

Advanced Materials in Nanofiltration Membrane

3.1 Overview of Advanced Materials Used in Thin Film (Nano)Composite Membrane

Membrane science research can be generally divided into seven major areas, that is, material selection, material characterization, membrane fabrication, membrane characterization and evaluation, transport phenomena, membrane module design, and process performance. Among these areas, materials chosen for membrane fabrication are the most important part in membrane technology and this is reflected by the significant amount of technical articles published in the literature.

Many attempts have been made in the past 10–15 years to enhance separation performance of polymeric NF membrane through variation of different parameters involved during the membrane preparation process such as dope formulation and casting conditions (for substrate making), aqueous/organic solution properties, and IP conditions (for PA layer synthesis), etc. Of these parameters, it is found that the utilization of advanced materials in preparing membrane of improved properties still remains the top priority among the community of membrane scientists worldwide, whether in the past or present.

The focus of this chapter is to review the role of these advanced materials in preparing NF membrane as a sustainable technological solution to not only water/wastewater separation and purification processes but also applications in organic solvents. Owing to the importance and popularity of the NF membrane made via the IP process for industrial applications, the following sections will provide a review on the advanced materials used in the development of TFC/TFN membranes. Special focus and emphasis will be placed on the role of advanced materials in improving membrane properties with respect to permeability/selectivity, chlorine tolerance, solvent stability, and fouling resistance.

In order to facilitate better understanding of the content, this chapter is organized into two main sections, (a) advanced materials used in PA selective layer synthesis and (b) advanced materials used in microporous substrate making. Because of the similarities of membrane preparation method and structural properties, the RO membrane prepared via IP between two different active monomers will also be considered here, although the composite NF membrane in general exhibits lower monovalent salt rejection than that of the composite RO membrane.

3.2 Advanced Materials in PA Thin Layer

3.2.1 Monomer

As composite membrane performances are mainly determined by film structure (i.e., pore dimension, thickness, roughness, and hydrophilicity) and its chemical properties (i.e., CL, functional groups, and bonds), a fundamental understanding of the effects of different monomers on composite membrane properties is necessary in order to tailor desired membrane structure and separation performance. Table 3.1 shows the organic structures of commonly used monomers while Table 3.2 summarizes the novel monomers that have been used in thin film formation over the past decade. The commonly used reactive monomers are aliphatic/aromatic diamine such as piperazine (PIP), *m*-phenylenediamine (MPD), and *p*-phenylenediamine (PPD) and acyl chloride monomers such as TMC, isophthaloyl chloride (IPC), and 5-isocyanato-isophthaloyl chloride (ICIC). Among these materials, the cross-linked aromatic PA composite membrane produced by IP of MPD and TMC is the most successful commercial product. Typically, an interfacially polymerized layer is prepared from CL between amine and acyl chloride monomers on a microporous substrate that is made of either PSF or PES. In order to enhance the mechanical properties of the composite membrane that is used in the high-pressure filtration process, nonwoven polyester fabric is often used as support to the microporous substrate.

Despite the significant high flux of the composite membrane compared to the asymmetric membrane in water separation processes, the research on how to further improve the properties of the top active skin layer still remains a high-priority domain over the years. Table 3.3 compares the optimum performances of selected TFC NF membranes prepared from novel monomers with commercially available NF membranes. By taking into consideration the applied operating pressure, one can realize that the water permeability coefficient of the newly developed TFC NF membranes (2.0–12.5 L m^{-2} h^{-1} bar^{-1}) is obviously lower than that of commercial NF270 membrane (20.5 L m^{-2} h^{-1} bar^{-1}). Even though some of the novel membranes demonstrated relatively lower water permeation rate, they on the other hand showed signs of improvement with respect to chlorine tolerance, solvent/pH stability, fouling resistance, etc. The following is the highlight of the research

TABLE 3.1

Organic Structure of Commonly Used Monomers for PA Thin Layer Synthesis Together with Their Respective MW

Amine Monomer (Abbreviation)	Chemical Structure	Molecular Weight (g mol^{-1})	Acyl Chloride Monomer (Abbreviation)	Chemical Structure	Molecular Weight (g mol^{-1})
Piperazine (PIP)	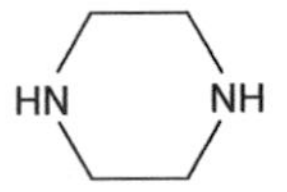 HN NH	86.14	Trimesoyl chloride (TMC)	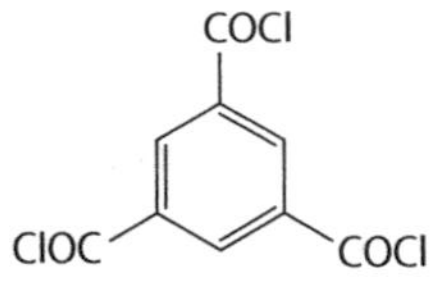 COCl, ClOC, COCl	265.48
m-Phenylenediamine (MPD)	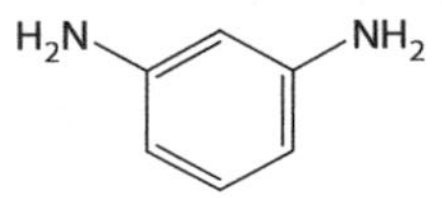 H_2N, NH_2	108.10	Isophthaloyl chloride (IPC)	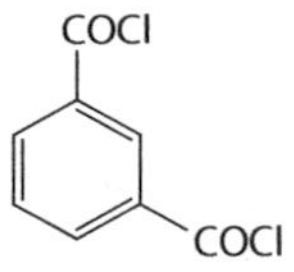 COCl, COCl	203.02
p-phenylenediamine (PPD)	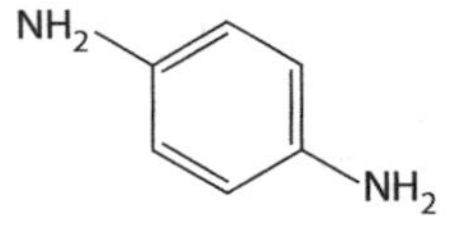 NH_2, NH_2	108.10	5-isocyanato-isophthaloyl chloride (ICIC)	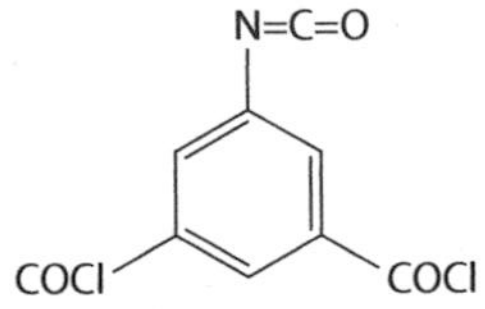 N=C=O, COCl, COCl	243.04

TABLE 3.2

Summary of Novel Monomers Used in TFC Membrane Preparation

Amine Monomer (Abbreviation)	Chemical Structure	Molecular Weight (g mol^{-1})	Acyl Chloride Monomer (Abbreviation)	Chemical Structure	Molecular Weight (g mol^{-1})
Sulfonated cardo poly(arylene ether sulfone) (SPES-NH_2)[a] Chen et al. (2008)		774.71	*mm*-biphenyl tetraacyl chloride (*mm*-BTEC)[b] Li et al. (2008), Wang et al. (2011)		404.03
3,5-diamino-N-(4-aminophenyl) benzamide (DABA)[a] Wang et al. (2010)		242.27	*om*-biphenyl tetraacyl chloride (*om*-BTEC)[b] Li et al. (2008)		404.03
Triethanolamine (TEOA)[b] Tang et al. (2008)		149.19	*op*-biphenyl tetraacyl chloride (*op*-BTEC)[b] Li et al. (2008)		404.03

(*Continued*)

TABLE 3.2 (*Continued*)

Summary of Novel Monomers Used in TFC Membrane Preparation

Amine Monomer (Abbreviation)	Chemical Structure	Molecular Weight (g mol^{-1})	Acyl Chloride Monomer (Abbreviation)	Chemical Structure	Molecular Weight (g mol^{-1})
Methyl-diethanolamine (MDEOA)[b] Tang et al. (2010)		119.16	Cyclohexane-1,3,5-tricarbonyl chloride (HTC)[a] Yu et al. (2009)		271.53
1,3-cyclohexanebis (methylamine) (CHMA)[a] Buch et al. (2008)		142.24	5-chloroformyloxy-isophthaloyl chloride (CFIC)[a] Liu et al. (2009)		281.48
m-phenylenediamine-4-methyl (MMPD)[a] Yu et al. (2009)		122.17	Naphthalene-1,3,6-trisulfonylchloride (NTSC)[b] Liu et al. (2012a)		423.35
Hexafluoroalcohol-m-phenylenediamine (HFA-MPD)[a] La et al. (2010)		530.31	Cyanuric chloride (CC)[b] Lee et al. (2015)		184.41

(Continued)

TABLE 3.2 (*Continued*)

Summary of Novel Monomers Used in TFC Membrane Preparation

Amine Monomer (Abbreviation)	Chemical Structure	Molecular Weight (g mol^{-1})	Acyl Chloride Monomer (Abbreviation)	Chemical Structure	Molecular Weight (g mol^{-1})
Polyvinylamine (PVAm)[b] Yu et al. (2011), Liu et al. (2012c)	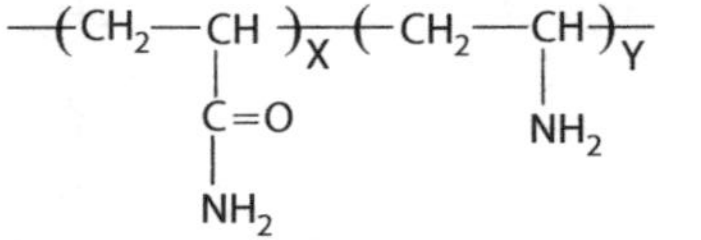 $-(CH_2-CH(C{=}O\,NH_2))_X-(CH_2-CH(NH_2))_Y-$	N/A			
Melamine[b] Han (2013)	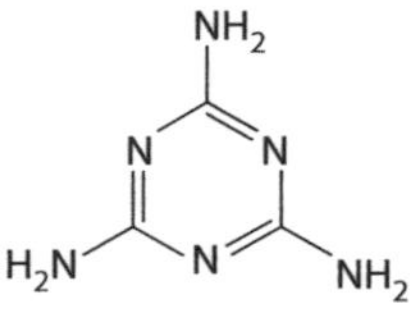 NH_2, H_2N, NH_2; N, N, N	126.15			
Diaminopiperazine (DAP)[b] Aburabie et al. (2015)	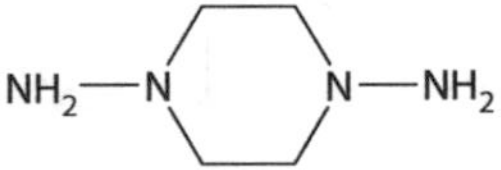 NH_2-N … $N-NH_2$	116.17			
1,4-bis(3-aminopropyl) piperazine (DAPP)[b] Li et al. (2015a)	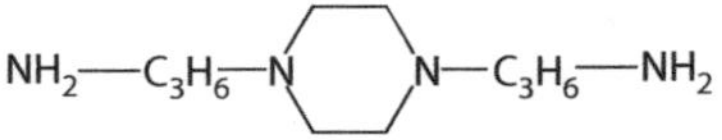 $NH_2-C_3H_6-N$ … $N-C_3H_6-NH_2$	200.33			
Dopamine[b] Zhao et al. (2014)	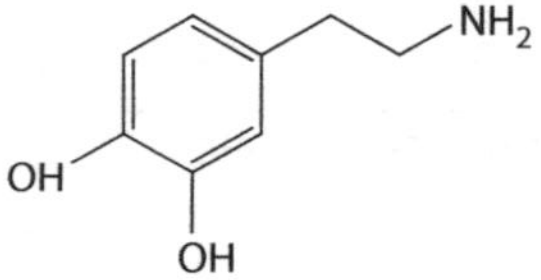 NH_2, OH, OH	153.18			

[a] Monomer used for PA RO membrane preparation.
[b] Monomer used for PA NF membrane preparation.

TABLE 3.3

Comparison between the Performances of Selected In-House Made TFC NF Membranes and Commercial NF Membranes

Amine Monomer	Acyl Chloride Monomer	Substrate[a]	Preparation Conditions[b]	Test Conditions	Optimum Performances[c]
CHMA	TMC	PES	Aqueous solution—0.5%–3% CHMA in deionized water (1 min) Organic solution—0.05%–0.35% TMC in toluene (10–90 s) Post treatment—oven at 55–70°C (3 min)	Pressure = 10 kg cm^{-2} (9.81 bar) Temperature = N/A Solution = 2000 ppm NaCl Filtration mode = N/A	J_{WF} = ~65 L m^{-2} h^{-1} R_{NaCl} = ~75% Buch et al. (2008)
Natural polymer sericin	TMC	PSF	Aqueous solution—0.5% sericin, 0.005 SDS% in aqueous NaOH solution of pH 9.5 (3 min) Organic solution—0.06% TMC in Isopar G (60 s) Post treatment—oven at 70°C (15 min)	Pressure = 0.5 MPa (5 bar) Temperature = 25°C Solution = 500 ppm Na_2SO_4 Filtration mode = Cross flow	J_{PWF} = 11.9 L m^{-2} h^{-1} bar^{-1} $R_{Na_2SO_4}$ = 95.3% Zhou et al. (2014)
BPA	TMC	PES	Aqueous solution—0.1%–2% BPA in aqueous NaOH solution of pH 11 (15 min) Organic solution—0.15% TMC in hexane (10–60 s) Post treatment—dried in air (30 min)	Pressure = 6 × 10^5 Pa (6 bar) Temperature = N/A Solution = 15 mg/L humic acid (pH3) Filtration mode = Cross flow Cross flow velocity = 0.4 L min^{-1}	J_{WF} = ~2.88 L m^{-2} h^{-1} bar^{-1} FR_w = 15% Abu Seman et al. (2010)
PEI	CC	PESf	Aqueous solution—1.25 g L^{-1} PEI in water (10 min) Organic solution—0.5 g L^{-1} CC in hexane (30 s) Post treatment—N/A	Pressure = 10 bar Temperature = Room temperature Solution = 2000 ppm NaCl or $MgSO_4$ Filtration mode = Dead end	J_{WF}= ~2 L m^{-2} h^{-1} bar^{-1} R_{NaCl} = ~45% $R_{Na_2SO_4}$ = ~68% Lee et al. (2015)

(Continued)

TABLE 3.3 (*Continued*)

Comparison between the Performances of Selected In-House Made TFC NF Membranes and Commercial NF Membranes

Amine Monomer	Acyl Chloride Monomer	Substrate[a]	Preparation Conditions[b]	Test Conditions	Optimum Performances[c]
PIP	NTSC	PSF	Aqueous solution—0.08%–2.64% PIP, 4% TEA, and 0.01% SDS in DI water (5 min) Organic solution—0.028%–0.032% NTSC and 0%–1.5% ethylene glycol monomethyl ether in Isopar G (8 min) Post treatment—dryer at 95°C (10 min)	Pressure = 0.5 MPa (5 bar) Temperature = 25°C Solution = 500 ppm Na_2SO_4 Filtration mode = Cross flow	J_{WF} = 5.76 L m^{-2} h^{-1} bar^{-1} $R_{Na_2SO_4}$ = 86.7% Liu et al. (2012a)
PVAm	IPC	PSF	Aqueous solution—0.4%–1.0% PVAm, 6% TEA, and 0.05% SDS in DI water (5 min) Organic solution—0.04%–0.12% IPC in Isopar G (1 min) Post treatment—dryer at 80°C (10 min)	Pressure = 0.6 MPa (6 bar) Temperature = 25°C Solution = 100 ppm NaCl or $MgSO_4$ Filtration mode = Cross flow	J_{PWP} = 23.8 L m^{-2} h^{-1} R_{MgSO_4} = 92.1% R_{NaCl} = 61.3% Yu et al. (2011)
MPD[d]	TMC	PSF	*Unknown* (proprietary)	Pressure = 9 bar Temperature = 25°C Solution = 1000 ppm NaCl or Na_2SO_4 Filtration mode = Dead end	J_{PWP} = 5.4 L m^{-2} h^{-1} bar^{-1} R_{NaCl} = 50.2% $R_{Na_2SO_4}$ = 98.3% Ong et al. (2012)
PIP[e]	TMC	PSF	*Unknown* (proprietary)	Pressure = 9 bar Temperature = 25°C Solution = 1000 ppm NaCl or Na_2SO_4 Filtration mode = Dead end	J_{PWP} = 20.5 L m^{-2} h^{-1} bar^{-1} R_{NaCl} = 41.2% $R_{Na_2SO_4}$ = 82.2% Ong et al. (2012)

[a] PES, polyethersulfone; PPSU, polyphenylsulfone; PAN, polyacrylonitrile; PSF, polysulfone; PESf, hydrophilized polyethersulfone.

[b] The subscript number indicates the sequence of process while the time (in bracket) shows the immersion/process period. All the components added were calculated based on w/v%, unless stated otherwise.

[c] J_{PWP} = pure water flux, J_{WF} = water flux, FR_w = irreversible fouling factor, R = rejection.

[d] NF90 commercial membrane manufactured by Dow Filmtech™.

[e] NF270 commercial membrane manufactured by Dow Filmtech™.

outcomes achieved by researchers with the use of novel monomers over the past 10 years.

In 2008, Zhang and his coworkers (Li et al., 2008) synthesized a series of isomeric biphenyl tetraacyl chloride (BTEC) for the preparation of TFC membrane with MPD as amine monomer in aqueous solution. It is claimed that the organic phase reactant is likely to have a greater impact on membrane performance compared to aqueous phase reactant because the IP process is generally diffusion controlled in the organic layer. The experimental results showed that the membrane prepared from *op*-BTEC demonstrated the highest permeability (54.2 L m^{-2} h^{-1}) followed by membranes prepared from *om*-BTEC (50.0 L m^{-2} h^{-1}) and *mm*-BTEC (31.7 L m^{-2} h^{-1}) when tested using 2000 ppm NaCl solution at 2 MPa. The reason for the flux enhancement might be due to the rougher and larger surface area of the *op*-BTEC membrane produced which led to greater contact with water molecules. Very interestingly, the *op*-BTEC membrane did not suffer from a drawback of the trade-off between permeability and selectivity as its salt rejection remained almost unchanged. As a continuation of this work, Zhang and his coworkers further applied *mm*-BTEC as monomer in the preparation of NF membranes with different surface charges (Li et al., 2009; Wang et al., 2011). It is reported that the surface charge of NF prepared by IP of *mm*-BTEC and PIP could be easily manipulated through the use of different properties of organic solvent in which cyclohexane tended to produce negatively charged membrane while toluene (aromatic) tended to induce positive charge to the membrane.

Similar to the work led by Zhang (Li et al., 2008), Li et al. (2014) prepared TFC NF membranes using three types of aliphatic amine monomers of very similar structure, that is, diethylenetriamine (DETA), triethylenetetramine (TETA), and tetraethylenepentamine (TEPA). The properties and performance of these newly synthesized membranes were compared with the control PIP/TMC membrane. Unfortunately, none of them were comparable with the control membrane. Besides exhibiting higher rejection against Na_2SO_4 and $MgSO_4$, the water flux of the control membrane (75 L m^{-2} h^{-1} MPa^{-1}) was significantly higher than those of newly prepared membranes (33.5–51.0 L m^{-2} h^{-1} MPa^{-1}). The better filtration performance of the control membrane can be attributed to its lower surface contact angle and higher charge property.

Great improvements in TFC membranes were experienced by Chen et al. (2008) by incorporating water soluble amine reactants—sulfonated cardo poly(arylene ether sulfone) (SPES-NH_2) into aqueous solution containing MPD. Under the optimum preparation conditions, the TFC membrane prepared from SPES-NH_2 showed remarkable increase in water permeability (51.2 L m^{-2} h^{-1}) with slight decrease in salt rejection compared to the membrane prepared without SPES-NH_2 (37.4 L m^{-2} h^{-1}). The improved results are attributed to the incorporation of hydrophilic SPES-NH_2 to PA and/or higher degree of CL formed in the thin selective layer. In view of the importance of hydrophilicity on the TFC membrane performance, a novel amine monomer—3,5-diamino-N-(4-aminophenyl) benzamide (DABA) with three amino groups was synthesized and used together with diamines (MPD) in the TFC membrane fabrication (Wang et al., 2010). With increasing the DABA content in the aqueous phase from 0% to 0.25% (w/v), the membrane showed increase in water flux from 37.5 to 55.4 L m^{-2} h^{-1} while maintaining a high salt rejection (~98%) in filtering salt solution containing 2000 ppm NaCl at 2 MPa. Instrumental characterizations further revealed that the top membrane surface became more hydrophilic, smoother, and thinner as DABA concentration was increased in amine solution.

By manipulating the concentration of hyperbranched polyethyleneimine (HPEI) in aqueous solution, Sun et al. (2012) successfully synthesized a composite NF membrane with good balance of water flux and rejection using 2 wt% HPEI. This membrane displayed pure water permeability of 4.9 L m^{-2} h^{-1} bar^{-1} with $MgCl_2$ and Raffinose (MW: 504.4 g mol^{-1}) rejection recorded at 96.2% and 90.0%, respectively. In addition, the HPEI-TMC NF membrane was highly positively charged when the pH was below 9.0 (up to 90 mV) and demonstrated high removal rate against zwitterionic antibiotic cephalexin (MW: 365.4 g mol^{-1}) in pH ranging from 2 to 8. Li et al. (2015a) prepared a positively charged NF hollow fiber membrane using 1,4-bis(3-aminopropyl)piperazine (DAPP) and TMC as amide and acyl chloride monomer. DAPP has a similar structure to PIP except for the presence of two aminopropyl groups bounded to the N atoms of the ring (see Table 3.2). Results showed that the positively charged membrane achieved 7.9–8.4 L m^{-2} h^{-1} water flux in filtrating different types of inorganic salts with rejection order, $MgCl_2$ (70.4%) > $MgSO_4$ (36.0%) > NaCl (23.0%) > LiCl (21.8%) when tested at 3 bar using 2000 mg L^{-1} salt solution. Compared to the positively charged membrane reported in the work of Sun et al. (2012), the DAPP–TMC membrane has significantly lower charged values for pH below 8.5. This might be the main reason why the DAPP–TMC membrane has lower $MgCl_2$ rejection compared to the highly positive charged HPEI–TMC membrane.

An environmentally friendly yet economical material—triethanolamine (TEOA) was reported in the work of Tang et al. (2008) as the amine monomer to enhance the performance of the TFC NF membrane. It is of great interest to use the TEOA monomer because the tertiary amino group in this molecule can be flexibly transferred into the quaternary ammonium group through variation of feed pH. Moreover, this composite membrane is found to be particularly suitable for treating acidic solution because water flux increases with lowering feed pH due to the fact that the amino group on membrane surface can change into R_3HN^+, resulting in increased hydrophilicity in an acidic environment. Another type of novel NF membrane interfacially synthesized from polyvinlyamine (PVAm) and IPC is also found very suitable for treating acidic feed (Yu et al., 2011). The pure water permeability of the membrane was reported to increase by over 40% when the feed pH decreased from 6.6 to 3.0. At pH around isoelectric point (IEP) (pH range of 6.5–7.0) of the

PVAm–IPC membrane, it is explained that the membrane surface charge was apparently reduced which as a consequence resulted in shrinkage of pores and reduced membrane water flux. In addition to the TEOA and PVAm, an acid stable TFC membrane prepared from naphthalene-1,3,6-trisulfonylchloride (NTSC) and PIP was also reported by Liu et al. (2012a). This PIP–NTSC composite membrane showed minor change in separation performance and no evidence on the change of PA layer chemical properties after running with 4.9% (w/v) H_2SO_4 for 2 months. Furthermore, Lee et al. (2015) fabricated a novel TFC NF membrane that demonstrated exceptionally strong resistance toward nucleophilic attack induced by extreme pH conditions. Compared to the control TFC membrane made of polyetherimide (PEI) and TMC, the novel membrane that was derived from PEI and cyanuric chloride (CC) showed very stable permeate flux and rejection throughout 5-week exposure to pH 1 or pH 13. The excellent performance of the membrane in almost the entire pH range can be attributed to the unique PA chemistry that is formed without amide bonds. Nevertheless, this novel NF membrane was associated with low water permeability (average 2.5 L m^{-2} h^{-1} bar^{-1}) even though NaCl rejection could be consistently kept at approximately 60% regardless of pH value.

Besides using NF membranes for a salt removal process, Liu et al. (2012b) synthesized new NF membranes using novel sulfonated aromatic diamine monomers for a dye-containing wastewater treatment process. The composite membranes were prepared via IP between TMC and mixture of novel monomer/PIP. Under the optimum fabrication conditions, a membrane with good combination of water flux and color retention could be produced. When tested with 1000 ppm dye solution at 0.5 MPa, the 2,5-bis(4-amino-2-trifluoromethylphenoxy) benzenesulfonic acid (6FAPBS)/PIP–TMC composite membrane showed approximately 62 L m^{-2} h^{-1} water flux with around 78% and 85% rejection in retaining methyl orange (MW = 327.3 g mol^{-1}) and rhodamine B (MW = 479.0 g mol^{-1}), respectively. Further investigation showed that the 4,4′-bis(4-amino-2-trifluoromethylphenoxy) biphenyl-4,4′-disulfonic acid (6FBABDS)/PIP–TMC composite membrane (in the absence of 6FAPBS) could produce very similar water flux compared with 6FBABDS/6FAPBS/PIP–TMC membrane but demonstrated slightly higher color rejection. Rejection of methyl orange and rhodamine B was increased to 82% and 87%, respectively. It must be pointed out that the electrostatic repulsive interaction between dyes and membrane surface upon addition of the sulfonated aromatic diamine monomer (due to presence of sulfonic acid groups) during the IP process is an effective way to improve membrane performance in the dyeing wastewater treatment process.

A key limitation to commercial PA composite membranes is membrane degradation through contact with chlorine—one of the common disinfectants used in water and wastewater treatment. For this reason, concerted effort has been devoted to developing a novel PA chemical structure that possesses a high degree of resistance to chlorine attack. It is known that the changes in the chemical nature of PA upon exposure to chlorine can affect membrane performance, shortening its lifespan and increasing operating cost. In order to overcome this limitation, Buch et al. (2008) made an attempt to develop the chlorine stability of NF by IP of 1,3-cyclohexanebis(methylamine) (CHMA) in water with TMC in hexane under different conditions. The composite membrane with aromatic–cycloaliphatic PA top layer was then exposed to sodium hypochlorite (NaOCl)–NaCl mixed solution of various NaOCl concentrations to ascertain the impact of chlorine on the membrane properties. The composite membrane, however, failed to retain performances as both water flux and salt rejection decreased considerably upon chlorine exposure of 24 h and 1 ppm. It is elucidated that the conversion of amide N–H group to N–Cl group upon chlorine exposure is the main factor causing the hydrophobic character of the PA layer to increase, leading to remarkable flux deterioration. NF membrane made of a new combination of dopamine and TMC was also evaluated for its stability in NaOCl solution, but remarkable changes in water flux and rejection were recorded with increasing chlorine exposure from 0 to 3000 ppm-h (Zhao et al., 2014).

In comparison to the CHMA/TMC and polydopamine (PD)/TMC membrane, aromatic–cycloaliphatic PA membranes developed from *m*-phenylenediamine-4-methyl (MMPD) and 1,3,5-tricarbonyl chloride (HTC) showed attractive chlorine resistance of more than 3000 ppm-h Cl (Yu et al., 2009). The very significant improvement on the stability of the MMPD–HTC membrane is attributed to the use of an aromatic diamine compound with a mono CH_3 substituent at the *ortho* position that is believed can minimize the attack by chlorine present in the water (see Table 3.2). Apart from the high chlorine resistance, the MMPD–HTC membrane also exhibited greater water permeation following a higher degree of pendant group—COOH formed on PA skin layer, as shown in XPS results. In 2015, amino functional polyethylene glycol (PEG) was exploited as a potential amine monomer to improve membrane chlorine resistance (Chen et al., 2015). It is found that the absence of active H atom in the PA selective layer made of PEG600-NH_2/TMC and PEG-4arm-NH_2/TMC led the resultant NF membranes to have much higher chlorine stability than that of the control MPD/TMC membrane. Filtration data proved that there was little change in the water flux and $MgSO_4$ rejection of the two newly fabricated membranes upon immersion in 2000 ppm NaOCI solution for 24 h.

Liu et al. (2009) fabricated a series of TFC RO membranes through the IP of MPD with TMC, ICIC, and CFIC, separately in an effort to assess the influence of polyacyl chloride structure on the chlorine stability of the composite membranes prepared. The results indicated that the MPD–CFIC membrane possessed the highest degree of resistance against chlorine attack, followed by the MPD–TMC and MPD–ICIC membrane after hypochlorite exposure (up to 2500 ppm-h Cl) at pH 8.5. It is explained that it is much easier for the N-chlorination reaction to take place in the MPD–ICIC membrane due to the existence of the urea (–NHCONH–) bond and pendant group of –NHCOOH. In 2010, novel TFC membranes with high tolerance to chlorine were prepared via IP of high MW of hexafluoroalcohol (HFA)-substituted aromatic diamines and TMC (La et al., 2010). As HFA is an electron withdrawing group and sterically bulky, both electronic and steric factors could play a key role in protecting amide linkages and benzene rings against chlorine attack. Results obtained from nuclear magnetic resonance (NMR) spectroscope revealed that HFA-PA composite membrane suffered only minor changes in the spectrum after 17-h exposure of 500 ppm hypochlorous acid (HClO) at pH 5.5. In comparison, the control PA membrane was severely attacked by chlorine with more than 50% chlorination recorded after the chlorine treatment process, causing irreversible damage to the membrane structure. In this case, protons on the benzene ring of the control membrane are substituted by Cl via Orton rearrangement upon chlorine exposure. In 2013, Han for the first time utilized melamine as amine monomer to fabricate a poly(melamine/TMC) composite membrane. The findings showed that this membrane not only exhibited excellent thermal resistance (for temperature up to 95°C) but also showed no sign of reduction in Na_2SO_4 rejection even after 96-h immersion in chlorine solution (NaOCl, 200 ppm). The excellent performance of this novel composite membrane can be strongly linked to the conjugation structure (–NHCO– group) formed between the TMC and melamine of the triazine ring, leading to decrease in the reactivity of the N atom in the –NHCO– group.

PA membrane in general is found not suitable for the treatment of nonaqueous solutions. However, an increasing interest on the use of PA membrane for the organic solution process can be observed in the recent years. A general solvent stable composite membrane (including the substrate) is one that is non-swelling in many solvent categories. In 2006, a solvent stable composite hollow fiber membrane which was found stable in 70% ethanol solution over the 10-week studied period was successfully prepared from the monomer system of PEI and IPC (Korikov et al., 2006). In order to improve adhesion between the selective layer and the support (lumen side of hollow fibers), the organic monomer solution was held within the fiber for up to 10 min instead of passing the solution through the fiber. The action was to create a condition similar to that of the coating on flat films. Prior to this work on the PEI/IPC monomer system, the reaction of PEI and a diisocyanate was also found useful in the production of a stable composite membrane for polar aprotic solvents. Further description on this work can be found in the US patent filed by Black (1991). Dopamine was first explored by Zhao et al. (2014) as a monomer in aqueous solution to improve the structural stability of NF membrane in alcohol solution. After 12-day immersion in pure ethanol, the water flux of the membrane was reported to increase by 28.5% while rejection against Orange GII (MW: 452 g mol^{-1}) dropped from around 85% to approximately 81%. Compared to the control PIP/TMC membrane in which flux and orange GII rejection were experienced to drop by 30% and 50%, respectively after only 1-day exposure to alcohol (Peng et al., 2013), the improvement in the alcohol stability of the PD membrane was significant. This is likely due to the existence of numerous π–π and hydrogen bonding interactions which enhances adhesion between the active layer and porous substrates.

In 2015, Aburabie et al. fabricated a solvent resistant nanofiltration (SRNF) membrane via the IP of diaminopiperazine (DAP) and TMC over a cross-linked polythiosemicarbazide (PTSC) substrate. Although the cross-linked PTSC substrate has been demonstrated for its stability in several organic solvents such as dimenthylsulfoxide (DMSO), tetrahydrofuran (THF), and dimenthylformamide (DMF), there was no discussion given by the researchers on why the DAP having a similar structure to PIP could enhance the PA layer stability against the solvent. In addition, the running period of the DAP/TMC membrane in the solvent experiment was missing, although simple solvent flux and rejection data were given. Currently, the main research focus on this topic is to explore the potential of nanofillers in the formation of a novel PA–inorganic nanocomposite layer that is able to tackle the problem of extremely low solvent flux encountered by the bare PA layer. More discussion on this topic can be found in Section 3.2.3.

With respect to fouling resistance, Abu Seman et al. (2011) reported that a TFC membrane with improved antifouling properties could be developed by means of IP between bisphenol A (BPA) and TMC. Research has been continuously conducted for PA composite membrane development since the 1970s, but a composite membrane with a highly charged surface seems to be necessary to overcome one of the limitations of composite membrane, that is, fouling propensity. The highly negative charge of composite membrane coupled with a uniform top PA layer was reported as demonstrating a lower fouling factor against negative charge of humic acid. Besides BPA, natural polymer sericin and 2,2′-oxybis-ethylamine (2,2′-OEL) were also used as an aqueous monomer to prepare NF membranes with improved antifouling property against bovine serum albumin (BSA) (Zhou et al., 2014; Jin et al., 2015). The membrane made of sericin and TMC was reported to exhibit better antifouling property compared to the commercial NF270 membrane (DOW FILMTEC™) owing to its higher negative surface charge which created higher electrostatic repulsion between foulants and the membrane surface, resulting in lesser adsorption of foulants on the membrane surface (Zhou et al., 2014). Improvement in the membrane surface hydrophilicity upon addition of hydrophilic 2,2′-OEL into PIP aqueous solution was also considered as one main factor increasing the membrane fouling resistance as the hydrophilic surface tended to strengthen its binding to water and reduce protein adsorption, making the membrane easy to clean and achieving a higher flux recovery rate (Jin et al., 2015). As flux degradation due to the fouling problem is unavoidable in the composite membranes of water separation processes, the control of membrane fouling will undoubtedly remain a high-priority research domain in the years to come. Other state-of-the art approaches to tackle the fouling phenomena of TFC membrane will be further discussed in following section where antifouling materials such as silver salts and silver nanoparticles are considered.

3.2.2 Surfactant/Additive

Studies on the use of surfactants as additives for asymmetric membrane making have been previously conducted for gas separation and pervaporation processes. Compared to its use in TFC membrane fabrication, only a limited number of articles reporting the impacts of the surfactants on TFC membrane performance are available. As a surfactant is capable of altering the polymerization efficiency of the PA layer formation by helping the monomer in the water phase move into the organic layer, improved property of the composite membrane can thus be produced. In certain cases, surfactant is added into amine aqueous solution to improve wettability of substrate surface so that a greater efficiency of polymerization can take place (Duan et al., 2010).

In the study of Jegal et al. (2002), three different types of surfactants were used for synthesizing the PA layer. Among the surfactants used, it was found that only triethyl benzyl ammonium bromide (TEBAB) could enhance the properties of the resultant composite membrane, achieving 40% increase in the water flux upon 0.2% (w/v) TEBAB added. The use of other two surfactants—trimethyl benzyl ammonium bromide (TMBAB) and triethyl benzyl ammonium chloride (TEBAC) played no role in modifying interfacial properties. This revealed that not all surfactants displayed positive features on membrane performance. On the other hand, Mansourpanah et al. (2009) examined the changes in the thin layer characteristics by employing different charge properties of surfactants in the organic phase. The results showed that the presence of anionic sodium dodecyl sulfate (SDS) in the organic solution could form defects and cracks on the thin layer surface, leading to more permeation and low rejection while cationic cetyl trimethyl ammonium bromide (CTAB) and nonionic Triton X-100 produced a denser skin layer, owing to increase in amine monomer diffusion across the interface toward organic phase. It is interesting to note that the overall performance of composite membranes containing surfactant was still superior compared with the membranes prepared without surfactant addition.

Mansourpanah et al. (2011) further reported the effect of SDS, CTAB, and Triton X-100 surfactants in the aqueous phase on the performance of composite membranes for cadmium (Cd) and lead (Pb) removal. Of the membranes prepared, only the membranes in the presence of CTAB and Triton X-100 showed improvement on heavy metal removal. The results showed that highest rejection of Cd and Pb could be obtained when 0.1% (w/v) CTAB or 0.5% (w/v) Triton X-100 was added during the IP process. Under the optimum preparation conditions, the thin layer containing CTAB displayed 61% Pb and 33% Cd rejection while the Triton X-100 contained PA layer showed 60% Pb and 32% Cd rejection. Addition of SDS on the membrane, however, decreased the separation performance as a result of free micelles formation. The formed micelles affected the PA layer integrity as cracks were detected in the thin layer structure.

Saha and Joshi (2009) on the other hand experienced a variation in the characteristics of resultant membranes by increasing the surfactant concentration of sodium lauryl sulfate (SLS) from 0.1 to 0.5 wt%. At low concentration of SLS added, the performance of the membrane remained unchanged. Further increasing the concentration from 0.125 to 0.5 wt% caused the NaCl rejection to decrease from 48% to 10% when the experiments were conducted at 1.03 MPa using 2000 ppm salt solution. No explanation was given by the researchers on the changes, but it is generally believed that the presence of surfactant plays a role in decreasing interfacial tension which facilitates the mass transfer of amine molecules to the organic phase to react with hydrochloride.

In addition to surfactant, the presence of additives in aqueous/organic solution during the IP process is also important in altering the structure of the film. Attention was devoted by Tang et al. (2010) to the preparation of NF membrane by adding inorganic LiBr into the aqueous phase solution during the IP process. The surface of the composite membrane became smoother as the LiBr content increased from 0% to 3% (w/v). A rougher surface, however, was observed at a greater amount of LiBr (i.e., 5% and 7%). With respect to membrane performance, asynchronous change in membrane water flux and salt rejection was experienced with increasing LiBr content. This can be caused by the unique interaction of LiBr with both alcohol amine and carbonyl of TMC. Li^+ ion is generally known to be able to interact with the hydroxyl oxygen atom of alcohol amine, increasing the density and reactivity of –OH groups in aqueous solution; as a result, a dense layer of composite membrane is formed. Formation of a loose surface layer is also possible following the interaction between Li^+ ion and carbonyl of TMC because the acyl chloride group of TMC is preferred to hydrolyze in the existence of Li^+ ion in aqueous solution.

Attempt has also been made by Abu Tarboush et al. (2008) to render the TFC membrane surface more hydrophilic by incorporating hydrophilic surface modifying macromolecules (LSMMs) onto the active PA layer. It was observed that the LSSMs synthesized during *in situ* polymerization could migrate toward the top air–polymer interface rendering the membrane hydrophilic and produced the best-performing composite membrane with improved flux stability compared to the LSMMs-free composite membrane. The improved stability is ascribed to the well-dispersed LSMMs in the PA layer that increases the mechanical strength of the thin film layer formed. With respect to salt rejection behavior over a studied period, this newly fabricated membrane always showed higher separation efficiency (~96%) than that of the LSMMs-free membrane (~91%) when tested using synthetic seawater solution at 5.5 MPa. As an extension of this study, Rana et al. (2011) incorporated silver salts simultaneously with LSMM into the PA layer to mitigate biofouling effects due to various foulants. The results obtained confirmed the antibiofouling effect of silver salts in which silver nitrate demonstrated the highest antimicrobial fouling intensity in desalination of salty solution, followed by silver lactate and silver citrate hydrate. On the other hand, ammonium salt which acted as a catalyst to enhance transportation of the amine monomer was utilized during the IP process (Xiang et al., 2014). Six types of ammonium salts with different molecular weights (59–242 g mol^{-1}) and structures (symmetrical and asymmetrical) were tested by adding them separately in PIP aqueous solution. It is found that increasing quaternary ammonium cationic size tended to achieve a higher efficiency of PIP transporting into organic solution, resulting in a rougher PA layer and better separation performance.

A new concept for synthesis of TFC NF membranes with controllable active layer thickness and effective "nanopores" has been introduced by Kong et al. (2010). Additional solvent—acetone was added into the organic phase with the aim of eliminating the great immiscibility gap between water and hexane, making the IP reaction zone controllable. Using this approach, it was possible to produce the composite membrane with extremely high flux (8.0×10^{-12} m^3 m^{-2} Pa^{-1} s^{-1}) (*Note:* 1 m^3 m^{-2} Pa^{-1} s^{-1} = 3.6 $\times$ 10^{11} L m^{-2} h^{-1} bar^{-1}) and no considerable salt rejection loss. Its water flux was remarkably higher compared with the membrane prepared by the conventional IP procedure (2.1×10^{-12} m^3 m^{-2} Pa^{-1} s^{-1}). It is also noted that with the appropriate addition of co-solvent content, a composite membrane with thinner denser layer and smoother top surface could be produced which is able to reduce water transport resistance and improve water flux. Using this so-called co-solvent assisted IP, Kong et al. (2011) further enhanced the water flux of PA membrane and achieved 1×10^{-11} m^3 m^{-2} Pa^{-1} s^{-1}. The best-performing membrane which was prepared by 2 wt% acetone addition also showed more than 99% rejection for glucose and maltose (500 ppm) and 95%–99% rejection for NaCl and $MgSO_4$ (2000 ppm) at operating pressure of 1.5 MPa.

In addition to acetone, remarkable flux enhancement on composite membrane was also reported elsewhere with the use of DMSO in the IP process (Kim, 2005). Similar to acetone, DMSO also plays influential role in reducing the solubility difference of two immiscible solutions. It increases penetration of diamine molecules into the organic phase, leading to the formation of a higher degree of CL PA layer. Aiming to produce a PA layer with tunable thickness, Liu et al. (2015) introduced different types of cyclic ethers, that is, oxolane, dioxane, and trioxane into the organic solution to increase the driving force of mass transport against PIP in hexane solution. The results showed that in comparison to the conventional IP process, the use of dioxane could create a narrower reaction zone that led to the formation of a thinner selective layer and improved membrane water flux. Although increasing the polarity of co-solvent from dioxane, oxolane to trioxane could gradually improve membrane water permeability, it also enlarged membrane pore dimension at the same time, affecting permeate quality. This is ascribed to a wider IP reaction zone that resulted from the increased miscibility of the interface.

In 2010, hexamethyl phosphoramide (HMPA) was first reported as an additive in aqueous solution for PA thin film formation (Duan et al., 2010). As HMPA is capable of facilitating the diffusion rate of the amine monomer into the organic phase, it may create a thicker zone of reaction. In comparison to the nonadditive membrane, the composite membrane prepared from 3 wt% HMPA showed dramatic increase in water flux (>70%) while maintaining a good rejection against salt. The substantial flux enhancement results from the combination effects of the enhanced hydrophilicity of the membrane surface coupled with the increased CL extent of the PA skin layer upon HMPA addition. By adding tributyl phosphate (TBP) and triphenyl phosphate (TPP) to an organic solution containing TMC, a high flux TFC membrane was reported (Kim et al., 2013). It is explained that the presence of TBP in the organic solution could considerably improve membrane water flux compared to that without TBP and with TPP. The membrane water flux was increased remarkably from around 40 gal ft^{-2} day^{-1} (*Note:* 1 gal ft^{-2} day^{-1} = 1.66 L m^{-2} h^{-1}) to around 68 gal ft^{-2} day^{-1} when the amount of TBP in isoparaffin was increased from 0 to 0.9 wt%. It must be pointed out that a significant increase in water flux did not cause a significant loss of salt rejection. Theoretically, it is known that the phosphate group of TBP is capable of forming "associations" with the carbonyl chloride group of TMC because of the dipole–dipole interaction between these two specific groups. The researchers of this work, however, did not provide a detailed discussion on how this dipole–dipole interaction could overcome the trade-off effect of membrane water flux and salt rejection.

3.2.3 Inorganic Nanomaterial

Recent advances in incorporating the organic PA layer with inorganic nanomaterials have further improved the bare PA layer characteristics, overcoming the trade-off effect between water permeability and solute selectivity. This new generation of composite membrane—TFN membrane was first reported by Hoek and his coworkers (Jeong et al., 2007) in their pioneering research work published in *Journal of Membrane Science* in early 2007. In this work, a new concept was introduced for the fabrication of novel RO membrane via the IP process by embedding zeolite NaA nanoparticles (in the range of 0.004%–0.4% (w/v)) within the PA layer made of MPD and TMC. A remarkable membrane flux enhancement was achieved upon zeolite nanoparticle incorporation with solute rejection remaining comparable to the typically prepared TFC membrane. It was suggested that water molecules tend to flow preferentially through the super-hydrophilic and molecular sieve nanoparticle pores.

Table 3.4 summarizes the characteristics of inorganic nanomaterials and their influences on the properties of TFN NF membranes upon incorporation. One of the earliest mentions of the TFN NF membrane was reported by Lee et al. in 2008, 1 year after the publication by Hoek and his coworkers

TABLE 3.4

Characteristics of Advanced Nanomaterials and Their Impacts on TFN NF Membranes Properties

Nanomaterial (Trade Name)	Formula/Structure	Properties	Preparation Conditions[a]	Effects on Membrane Properties
Solid silica (LUDOX® HS40) Jadav and Singh (2009)	SiO_2	Colloidal silica (40% suspension in water) Radius of gyration: 6.6 nm	Aqueous solution—2% MPD, 0.002%–0.5% SiO_2 in DI water (40 s) Organic solution—0.1% TMC in hexane (40 s) Post treatment—oven at 70°C (3 min)	• Water flux increased with increase in silica content in aqueous solution • Optimal salt rejection was obtained when aqueous solution containing 0.022% SiO_2
Modified mesoporous silica (self-synthesized) Wu et al. (2013)	SiO_2	Water affinity: Hydrophilic Particle size: 100 nm Pore size: 2.2 nm BET: 837 $m^2\ g^{-1}$	Aqueous solution—10 mg L^{-1} PIP, 2 mg L^{-1} NaOH, 0%–0.07% SiO_2 in DI water (5 min) Organic solution—1.5 mg L^{-1} TMC in cyclohexane (1 min) Post treatment—oven at 70°C (30 min)	• Good combination of flux and Na_2SO_4 rejection was achieved upon 0.03% SiO_2 incorporation • Both flux and rejection were affected with the excessive use of SiO_2 (0.05%–0.07%)
Hydrophilized ordered mesoporous carbon (H-OMC) (self-synthesized) Kim and Deng (2011)		Water affinity: Hydrophilic Mean pore diameter: 3.7 nm Interval between carbon rods: 1–3 nm	Aqueous solution—2% MPD, 0%–10% H-OMC in DI water (3 min) Organic solution—0.2% TMC in hexane (3 min) Post treatment—oven at 60–70°C (120 min)	• Increasing H-OMC loading (up to 6%) increased membrane hydrophilicity, wettability, and permeability • Rejection of Na_2SO_4 was decreased slightly with increasing nanofiller loading

(Continued)

TABLE 3.4 (*Continued*)

Characteristics of Advanced Nanomaterials and Their Impacts on TFN NF Membranes Properties

Nanomaterial (Trade Name)	Formula/Structure	Properties	Preparation Conditions[a]	Effects on Membrane Properties
Graphene oxide (GO) (Sigma-Aldrich) Xia et al. (2015)	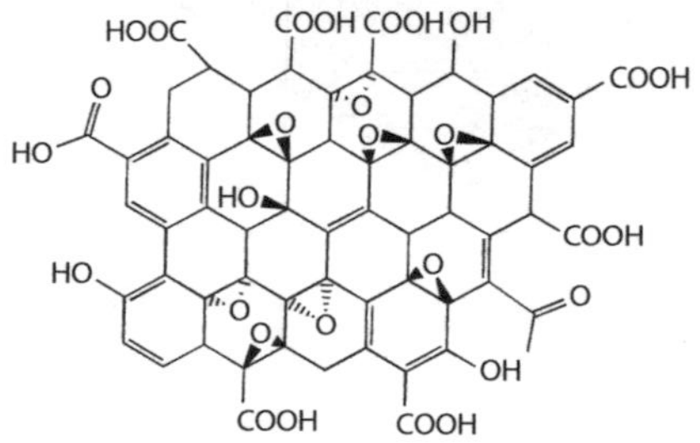	Water affinity: Superior hydrophilic Structure: Flake Charge: Negative	Aqueous solution—2% MPD, 0%–0.012% GO in DI water (2 min) Organic solution—0.1% TMC in hexane (1 min) Post treatment—oven at 60°C (8 min)	• Water flux and hydrophilicity (contact angle) increased by 24% and 10%, respectively, with increasing GO loading from 0% to 0.004% • Antifouling resistance improved upon GO incorporation
Zeolite–NaX (self-synthesized) Fathizadeh et al. (2011)	Synthesized from $NaAlO_2$	Water affinity: Hydrophilic Average crystal size: 112 nm Entrance pore: ~7.4 Å	Aqueous solution—2% MPD in DI water (2 min) Organic solution—0.1% TMC and 0.004%–0.2% NaX in hexane (1 min) Post treatment—oven at 70°C (6 min)	• Water permeate flux increased with increasing NaX content in hexane • Only slight variation in NaCl rejection was observed for all membranes prepared
Silver nanoparticle (Sigma-Aldrich) Lee et al. (2007)	Ag	N/A *Note: Specific properties of silver nanoparticles used were not provided*	Aqueous solution—2% MPD, 1.5% Ag in DI water (*unknown*) Organic solution—0.2% TMC in 1,1-dichloro-1-flueroethane (*unknown*) Post treatment—N/A (*unknown*)	• The presence of Ag in PA membrane has little influence on permeability and selectivity. However, the resultant membrane exhibited high resistance against biofouling

(*Continued*)

TABLE 3.4 (*Continued*)

Characteristics of Advanced Nanomaterials and Their Impacts on TFN NF Membranes Properties

Nanomaterial (Trade Name)	Formula/Structure	Properties	Preparation Conditions[a]	Effects on Membrane Properties
Single-wall carbon nanotube (SWCNT) (self-synthesized) Baroña et al. (2012)		Water affinity: Hydrophilic O.D: 2.3–2.7 nm I.D: 1 nm Length: At least 200 nm	Aqueous solution—1.5% PVA, nanotubes gel (5–10 mL added into 100 ml solution) in DI water (5 min) Cross-linker solution—100 mL water, 100 mL acetone, 1 g H_2SO_4 and 2.5 g of 25% glutaraldehyde solution (30 min) Post treatment—washed with ethanol and dried at 80°C (30 min)	• Hydrophilicity of membrane improved due to the decrease in contact angle, leading to greater water flux • SWCNT was compatible with polyvinyl alcohol layer as no large defects or cracks were observed.
MWCNT (Chengdu Organic Chemicals Co. Ltd., China) Wu et al. (2010)		O.D: <8 nm I.D: 2–5 nm Length: 10–30 μm Purity: >95% COOH content: 3.86 wt%	Organic solution—0.6% TMC in hexane (30 min) Aqueous solution—6% TEOA, 0.3% SDS, and 0.05% MWCNT in DI water (5–40 min) Organic solution—0.6% TMC in hexane (30 min) Post treatment—oven at 60°C (30 min)	• Pure water flux increased while contact angle decreased with increasing immersion time in aqueous solution • Membranes displayed low degree of CL and possessed high amount of carboxyl groups

[a] The subscript number indicates the sequence of process while the time (in bracket) shows the immersion/process period. All the components added were calculated based on w/v%, except otherwise stated.

(Jeong et al., 2007). Although the researchers did not name the membranes as TFN, the concept of embedding nanomaterials, that is, titanium dioxide (TiO_2) nanoparticles into the PA layer via IP was the same approach as that for TFN RO fabrication (Jeong et al., 2007). Instead of hexane, which is most frequently used in the organic phase, Lee et al. (2008) used 1,1-dichloro-1-fluroethane (HCFC) to better disperse high loadings of TiO_2 nanoparticles (up to 10 wt%). The results showed that the membrane flux declined slightly while $MgSO_4$ rejection increased with increasing TiO_2 loading from 0 to 5 wt%. Further increase in TiO_2 loadings to 10 wt%, however, resulted in an abrupt flux increase with extremely poor rejection (<5%). The poor separation performance was attributed to the weak mechanical strength of the membrane resulting from the easier peeling-off of the PA–TiO_2 layer from the substrate. In addition, the possibility of a decrease in the degree of PA CL resulting from the excessive use of nanoparticles was not ruled out as the presence of inorganic nanofillers might act as end groups to interfere with the polymerization reaction. Self-assembly of TiO_2 nanoparticles on the TFC membrane was also investigated to reduce membrane biofouling (Ho et al., 2003). It is claimed that the embedded TiO_2 is less likely to be washed away from the membrane surface during the filtration process, owing to the bidentate coordination of carboxylate to Ti^{4+} and H-bonding interaction between the –COOH functional group and TiO_2 nanoparticles. As TiO_2 could be hybridized with the membrane top structure by simply dipping the neat composite membrane in the TiO_2 colloidal solution, a new type of TFC membrane demonstrating great potential in microbial fouling prevention could be easily produced.

In an effort to minimize agglomeration of TiO_2 nanoparticles on the TFN NF membrane surface, Rajaeian et al. (2013) modified the particle surface using an aminosilane coupling agent—N-[3-(trimethoxysilyl)propyl] ethylenediamine (AAPTS), before the particle was used in IP. It was assumed by the authors that the presence of silane functional groups on the TiO_2 surface could effectively minimize the probability of Ti–O–Ti oxygen bridge bonds between unmodified TiO_2 nanoparticles and consequent particle agglomeration. The experimental results showed that all the TFN membranes made of the modified TiO_2 exhibited better water flux than the control TFC membrane in which the PWF increased progressively from 11.2 L m^{-2} h^{-1} of TFC membrane to approximately 13.2, approximately 20 and 27 L m^{-2} h^{-1} of TFN membranes, respectively, corresponding to the particle loading of 0.005, 0.05 and 0.1 wt%, when tested at approximately 7.5 bar (110 psi). NaCl rejection improved slightly to 54% when a small quantity of modified TiO_2 (i.e., 0.005 wt%) was incorporated into the membrane because of the increased binding affinity of particles to the PA network during the IP process. The improvement in salt rejection could not be maintained at the highest loading of modified TiO_2, mainly because of the decrease of CL in the PA layer. It is further elucidated that the IP reaction would be impaired by the amine functional groups grafted on the TiO_2 surface when high loading of modified

TiO_2 was added as the presence of amine groups on TiO_2 surface tended to compete with MPD for reaction with TMC.

Besides hydrophilic TiO_2 nanoparticles, a two-dimensional single layer material bearing *sp*^{2}-bonded carbon–graphene oxide (GO) has drawn the attention of scientists as one of the promising nanofillers in TFN membrane making. The existence of large numbers of covalent oxygen-containing functional groups such as carboxyl, hydroxyl, and epoxide in GO has made this material possess not only negative charge but also superior hydrophilic property. Compared to the control TFC membrane, the TFN membrane incorporated with an appropriate amount of GO (0.012 wt%) in the PA layer could offer not only higher water flux and natural organic matter removal rate, but also better antifouling properties (Xia et al., 2015). This great potential for water treatment process is ascribed to the favorable change in membrane hydrophilicity, surface morphology, and surface charge upon the addition of GO.

While Hu et al. (2012a) reported the potential use of solid SiO_2 nanoparticles (particle size: 11 nm) to improve the separation characteristics of TFN NF membrane, Wu et al. (2013) explored the possibility of incorporating mesoporous silica nanoparticles (MSN) having average pore size of 2.2 nm into the PA layer of PIP and TMC. In order to enhance the compatibility between MSN and the polymer matrix, 3-aminopropyltriethoxysilane (APS) was used to functionalize MSN with amino groups ($-NH_2$). Figure 3.1 shows the schematic preparation process of modified MSN (mMSN) and its transmission electron microscopy (TEM) images. The high-resolution TEM image further confirmed the ordered mesoporous network of mMSN which was in a hexagonal array. The findings of this work showed that remarkable flux improvement with only a slight drop in Na_2SO_4 rejection was achieved when only 0.03 wt% mMSN was incorporated within the PA layer of TFN membrane. The increase in water flux can be mainly attributed to the additional pathway created by mMSN for water molecules. The increase in the system free volume resulting from the disruption of the polymer chain packing upon nanoparticles addition is also considered as one of the factors leading to flux increment (Hu et al., 2012a). Similar to most of the TFN membranes discussed earlier, the excessive incorporation of mMSN (>0.03 wt%) within the TFN membrane would negatively affect salt rejection rate.

Kim and Deng (2011) synthesized hydrophilized ordered mesoporous carbons (H-OMCs) and impregnated them in a thin PA film by *in situ* IP. It was hypothesized that (1) synthesized H-OMCs could be better dispersed in the aqueous phase compared to unmodified OMCs and resulted in stronger interaction with PA and (2) the ordered pore layer of H-OMCs (median pore diameter: 3.7 nm) could provide nanoflow paths for water molecules improving membrane flux production, provided the amount of nanomaterials used is appropriate. The TEM images in Figure 3.2 showed the existence of mesopores in the silica template and OMC materials. With respect to permeation characteristics, the researchers emphasized the importance of having an appropriate amount of H-OMCs in the polymer to achieve the nanopore

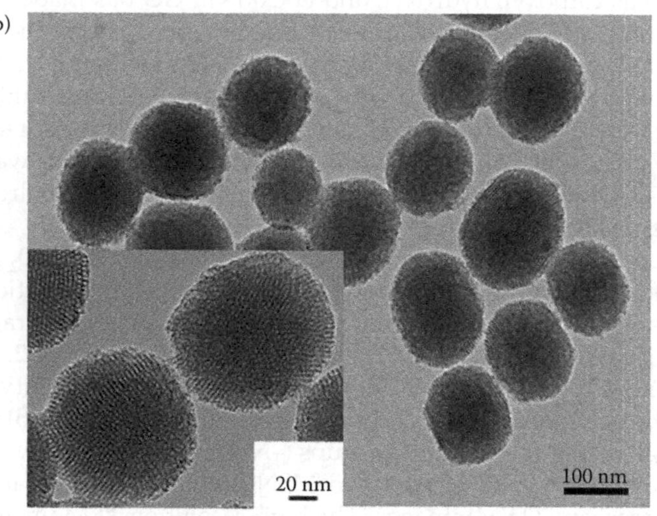

FIGURE 3.1
(a) Schematic preparation of mMSN and (b) TEM images of mMSN at two different scale bars. (Reprinted from *Journal of Membrane Science*, 428, Wu, H., Tang, B., Wu, P., Optimizing polyamide thin film composite membrane covalently bonded with modified mesoporous silica nanoparticles, 341–348, Copyright 2013, with permission from Elsevier.)

effect created by mesoporous carbon. Of the range of concentrations studied (0–10 wt%), the use of 5 wt% H-OMCs was the most ideal loading to produce the most permeable TFN membrane with NaCl and Na_2SO_4 rejection recorded at 48% and 88%, respectively.

The incorporation of MWCNTs throughout the superselective PA layer was also explored as a facile approach to produce superior hydrophilic composite membrane with fast water molecules transport (Wu et al., 2010). The unique molecular architecture of the tubes embedding in the membrane matrix has the potential to open up a new route to cheap, comparatively easy desalination. As the MWCNTs were not well dispersed in the nonpolar solvent of the organic phase, a modified IP process was proposed by immersing the support membrane into the organic phase first prior to the conventional IP process. The results obtained from TEM clearly showed a tube-like structure located within the cross section of thin film, indicating the presence of carbon nanotubes (CNTs) in the membrane. The TFN membrane made by the improved approach had shown to increase both permeability and

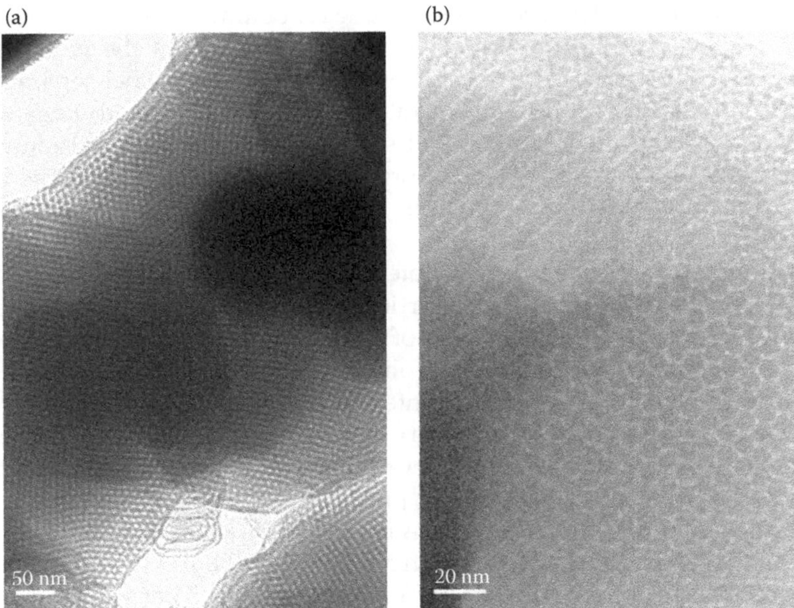

FIGURE 3.2
TEM images of (a) silica template (scale bar: 50 nm) and (b) ordered mesoporous carbon (OCM) sample (scale bar: 20 nm). (Reprinted from *Journal of Membrane Science*, 375, Kim, E.-S., Deng, B., Fabrication of polyamide thin-film nano-composite (PA-TFN) membrane with hydrophilized ordered mesoporous carbon (H-OMC) for water purifications, 46–54, Copyright 2011, with permission from Elsevier.)

selectivity (~4.5 L m^{-2} h^{-1} and ~78% Na_2SO_4 rejection at 0.6 MPa) when compared with the control TFC membrane (~1.6 L m^{-2} h^{-1} and ~70%) and the TFN membrane prepared by the conventional IP (~2.6 L m^{-2} h^{-1} and ~74%). It is explained that the promising separation is caused by the lower degree of CL thin layer produced coupled with the presence of a high amount of hydrophilic and negatively charged carboxyl groups on the top PA surface, as evidenced from XPS results. Besides hydrophilized CNTs, Roy et al. (2011) successfully functionalized MWCNTs with hydrophobic groups via microwave treatment and incorporated these hydrophobic CNTs into PA structure in order to create highly productive solvent resistant membranes for dye separation process in methanol solution. Baroña et al. (2012) also explored the possibility of embedding aluminosilicate single-walled nanotubes (SWNTs) into PVA/PSF composite NF membrane for salt water purification. Under the optimum loading of SWNTs (i.e., 20 v/v% in PVA solution), both water flux and Na_2SO_4 rejection of membrane were reported to increase without showing signs of the trade-off between permeability and selectivity. Rapid transport rate is likely to exist because nanotubes exhibit smoother walls (on atomic scales) than other materials and are of large surface area.

The application of the TFN NF membrane is not limited to aqueous media. There are some studies reporting on the performance of the membranes for nonaqueous separation process. Compared to conventional separations, solvent filtration using SRNF is an interesting alternative with benefits in terms of economy, environment, and safety. The pharmaceutical industry, for instance, uses a large amount of organic solvents as a reaction medium during the production and formulation of active pharmaceutical ingredients (APIs) (Geens et al., 2007).

An attempt was made to fabricate SRNF membranes by incorporating functionalized UZM-5 nanoparticles in the concentration range of 0–0.2 w/v% into the PA layer made of MPD and TMC (Namvar-Mahboub et al., 2014). These functionalized zeolites with an average particle size of 73 nm were claimed to be able to interact well with TMC monomers via covalent bonding to form a PA network as illustrated in Figure 3.3. However, the increase in organic phase viscosity coupled with an increase in miscibility between the aqueous and organic phases at higher zeolite loading led to the formation of a less cross-linked PA film. This, as a consequence, resulted in a gradual increase in solvent flux from 10.4 to 16.9 L m^{-2} h^{-1} with increasing zeolite loading from 0 to 0.2 w/v% when tested at 15 bar with feed solution consisting of 20 wt% lube oil, 40 wt% methyl ethyl ketone (MEK), and 40 wt% toluene. The increase in the solvent flux can be explained by the new transport pathways created by the functionalized UZM-5 for permeation of MEK and toluene. It was reported that the modified UZM-5 possessed an average pore diameter of 16.8 Å, that is, larger than the kinetic diameter of MEK (5.2 Å) and toluene (6.1 Å) but much smaller than those of the oil molecules. However, the separation efficiency of SRNF membrane against oil molecules was negatively affected when greater than 0.05 w/v% nanoparticles were embedded into the PA layer. This was considered to be due to significant agglomeration of nanoparticles, which was evidenced on the top surface of the membrane, causing the PA structure to distort and lowering its degree of CL.

On the other hand, an extraordinarily high methanol flux was reported by Peyravi et al. (2014) when functionalized TiO_2 nanoparticles were incorporated into the PA layer of SRNF membrane made of ethylenediamine and IPC. By incorporating nanoparticles functionalized by different compounds, that is, thionyl chloride (TCI), monoethanolamine (MEOA), and TETA, it was found that the resultant SRNF membranes could achieve superior methanol flux (120–140 L m^{-2} h^{-1}) when the membranes were tested at only 5 bar. In comparison to the performance of other SRNF membranes reported elsewhere such as commercial StarMem-120 and StarMem-188 (Geens et al., 2007), high flux TFC membrane (Jimenez-Solomon et al., 2012), in-house made TFC membrane (Kosaraju and Sirkar, 2008), and mixed matrix membrane (Sani et al., 2015), the methanol flux reported by Peyravi et al. (2014) was very high for a membrane used in solvent medium. The latter membranes demonstrated methanol flux at 3.6–5.5 L m^{-2} h^{-1} bar^{-1}, 7.0–46.0 L m^{-2} h^{-1} (at 30 bar),

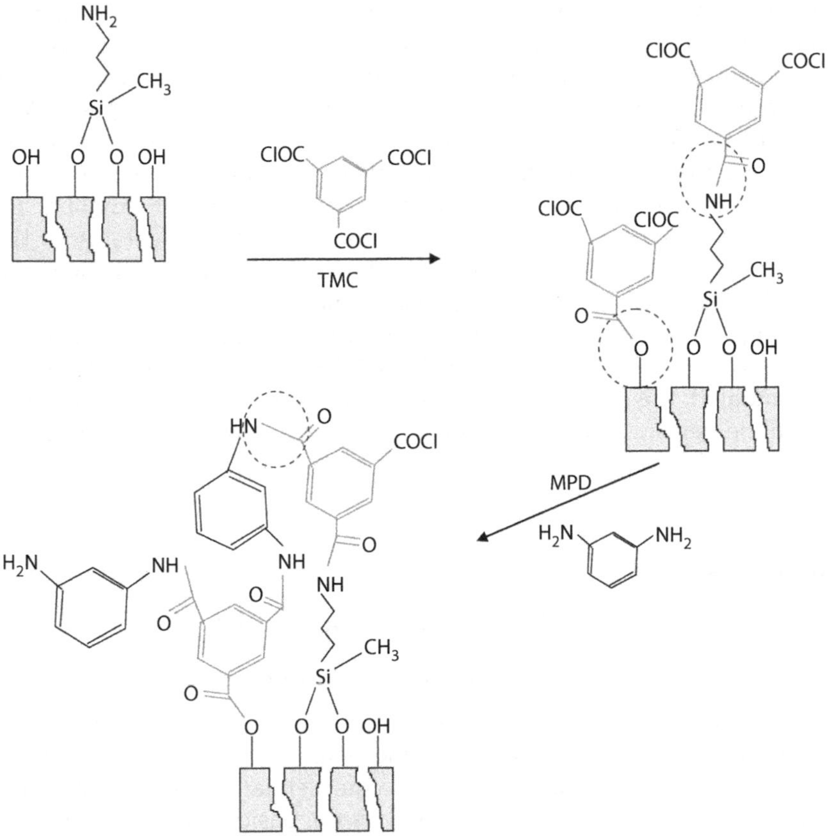

FIGURE 3.3
Schematic of chemical interaction between functionalized UZM-5 nanoparticles and TMC monomers during IP. (Reprinted from *Journal of Membrane Science*, 459, Namvar-Mahboub, M., Pakizeh, M., Davari, S., Preparation and characterization of UZM-5/polyamide thin film nanocomposite membrane for dewaxing solvent recovery, 22–32, Copyright 2014, with permission from Elsevier.)

approximately 6.41 L m^{-2} h^{-1} (at 4 bar), and approximately 73–100 L m^{-2} h^{-1} (at 10 bar), respectively. No explanation was provided by the researchers as to why the functionalized TiO_2 could be so powerful in improving the methanol flux of the TFN membrane by many orders of magnitude and could still maintain good selectivity, and no other groups have reported flux of similar magnitude. Typically, the significant low organic solvent flux of the membrane (compared to water flux) is mainly due to the high difference in surface tension between the solvent and membrane material that increases transport resistance of the solvent and reduces its permeation rate.

Although Basu et al. (2009) have also incorporated nanomaterials—metal-organic frameworks (MOFs) to prepare SRNF membrane, their method was different from the work of Namvar-Mahboub et al. (2014) and Peyravi et al. (2014). Instead of employing IP, a typical coating method was used to form a thin polydimethlysiloxane (PDMS) layer containing MOFs on the surface of microporous polyimide (PI) substrate. Of the four different types of MOFs studied ($Cu_3(BTC)_2$, MIL-47, MIL-53(Al), and ZIF-8), none of the SRNF membranes prepared could achieve solvent flux greater than 1.0 L m^{-2} h^{-1} bar^{-1} when tested using isopropanol solution containing 17.5 μmol L^{-1} Rose Bengal (MW: 974 g mol^{-1}). The extremely low solvent permeability of these membranes could be mainly due to the formation of a thick PDMS dense layer (30–35 μm) resulting from the coating method.

Although experiences with solvent resistant TFN NF membrane are still scarce and a successful application of TFN NF membrane in industry has not yet eventuated, the research of TFN membranes for aqueous medium filtration will be helpful to the progress of solvent resistant TFN NF membrane development in the future.

3.3 Advanced Materials in Microporous Substrate

3.3.1 Polymer/Polymer–Polymer-Based Substrate

A vast variety of polymers have been successfully used as porous supports for TFC membrane fabrication over the years. Among all the polymers used, PSF still remains a mainstay in commercial composite membrane till now. PSF in general is well accepted as the basic material for microporous substrate making, but it is not without its drawbacks being composite membrane support. The performance of PSF membrane could hardly be maintained at a temperature higher than 50°C in long run. Moreover, its hydrophobic nature and poor solvent stability are the other main concerns when it is used as a substrate of composite membrane in a solvent-based separation process (Petersen, 1993). This membrane generally fails to maintain its physical integrity as PSF-based substrate tends to swell or dissolve in organic solvents.

In order to enhance the solvent resistance of the membrane, Peyravi et al. (2012) modified the properties of the PSF substrate by introducing sulfonated poly(ether sulphide sulfone) (SPESS) as a secondary polymer during PSF dope preparation. The results showed that the swelling degree of the modified PSF TFC membrane in methanol solution was significantly lower than that of the TFC membrane made of unmodified PSF. The improvement in modified PSF with respect to solvent resistance as explained by the authors is mainly due to the formation of rigid ionic domain of SPESS in PSF matrix, hindering PSF chains to fluctuate freely in methanol. Table 3.5 summarizes the properties of polymers that have ever been used for making TFC membrane substrate. Very similar to the PA active layer development, studies

TABLE 3.5

Overview of the Properties of Polymeric Substrate Used for TFC Membrane Preparation

Polymer[a]	Structure Formula	T_g (°C)[b]	MW (g mol^{-1})[c]	Polymer (wt%) in Dope Solution (Configuration)	Characteristics
PSF (Udel P3500LCD, Solvay Adv. Polym.)	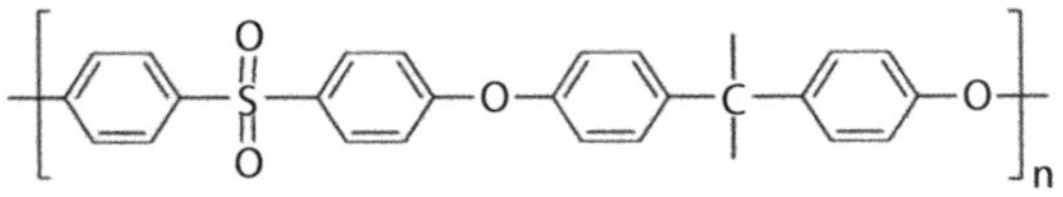	185	26,000	16.5–18 (flat sheet) 20 (hollow fiber)	Good thermal stability and high strength. Relatively low cost. Ghosh and Hoek (2009), Wang et al. (2010), Maurya et al. (2012)
PES (Ultrason E6020P, BASF)	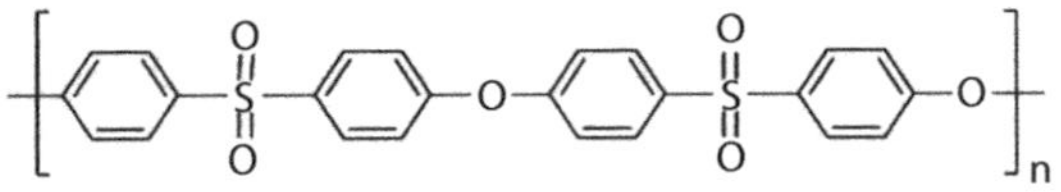	230[d]	58,000	16 (flat sheet)	Good thermal stability and high strength. Low resistance to hydrolysis which precludes the use of harsh conditions for cleaning process. Relatively low cost. Mansourpanah et al. (2009), Pourjafar et al. (2012)

(Continued)

TABLE 3.5 (*Continued*)

Overview of the Properties of Polymeric Substrate Used for TFC Membrane Preparation

Polymer[a]	Structure Formula	T_g (°C)[b]	MW (g mol^{-1})[c]	Polymer (wt%) in Dope Solution (Configuration)	Characteristics
PPSU (Radel®R5000, Solvay Adv. Polym.)		220	50,000	16 (flat sheet)	Better properties than PSF and PES. More resistant to hydrolysis and to plasticization by a number of organic solvents. Liu et al. (2012b)
PAN (GKSS-Forschungszentrum Geesthacht)		120[d]	N/A	N/A (flat sheet)	Relatively hydrophobic. Reasonable chemical stability. Dalwani et al. (2011)
PVDF (GE Osmonics)		–40/–38	N/A	N/A (flat sheet)	Good thermal and chemical resistance. Highly hydrophobic. Kim et al. (2009)

(*Continued*)

TABLE 3.5 (*Continued*)

Overview of the Properties of Polymeric Substrate Used for TFC Membrane Preparation

Polymer[a]	Structure Formula	T_g (°C)[b]	MW (g mol^{-1})[c]	Polymer (wt%) in Dope Solution (Configuration)	Characteristics
PP (Celgard, Charlotte, North Carolina)	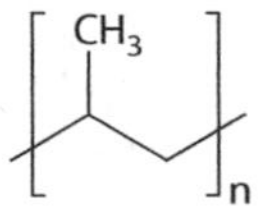	2/10	N/A	N/A (hollow fiber and flat sheet)	Extremely hydrophobic. High durability and resistance to chemicals, pH variations, and wide range of solvents. Korikov et al. (2006)
PEI (ULTEM™, Sabic)	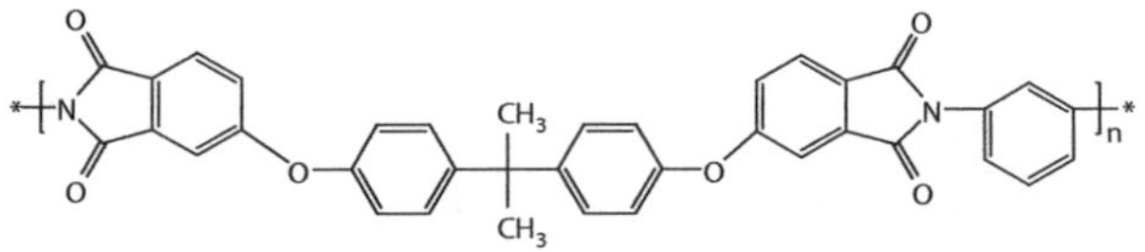	217	N/A	15 (flat sheet)	Hydrophobic. Excellent stability of physical and mechanical properties at elevated temperatures. Misdan et al. (2014)
PPBES (Dalian Polymer New Material Co. Ltd., China)	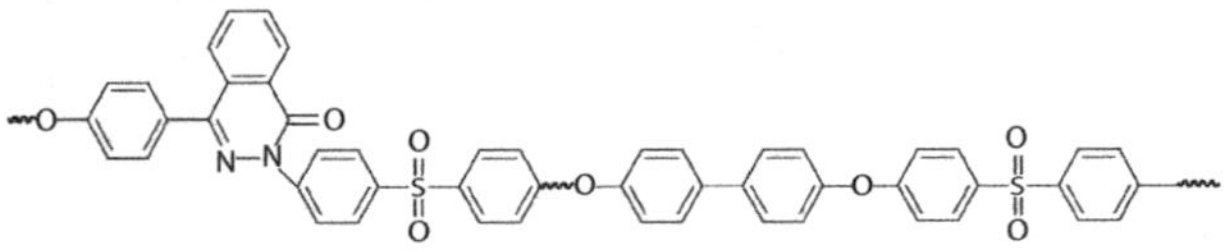	289	N/A	N/A	Better thermal resistance than support made of PPESK. Han et al. (2011)

(*Continued*)

TABLE 3.5 (*Continued*)

Overview of the Properties of Polymeric Substrate Used for TFC Membrane Preparation

Polymer[a]	Structure Formula	T_g (°C)[b]	MW (g mol^{-1})[c]	Polymer (wt%) in Dope Solution (Configuration)	Characteristics
PPESK (Dalian Polymer New Material Co. Ltd., China)	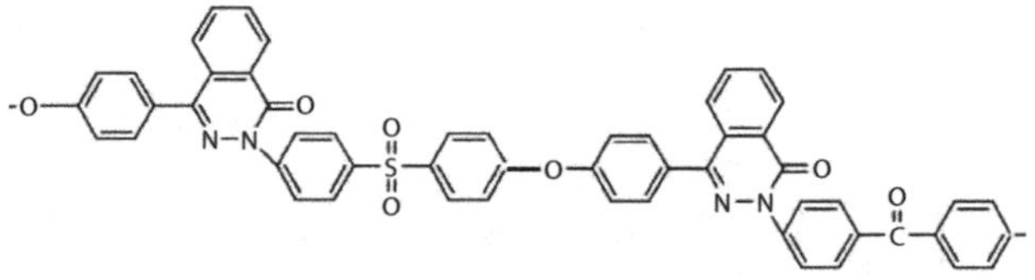	284	N/A	20 (hollow fiber)	Superior mechanical strength, thermal and chemical stability, and very high T_g. The PPESK membrane support, however, is more brittle than PES and PSF in common use. Yang et al. (2007)
PPENK (Dalian Polymer New Material Co. Ltd., China)	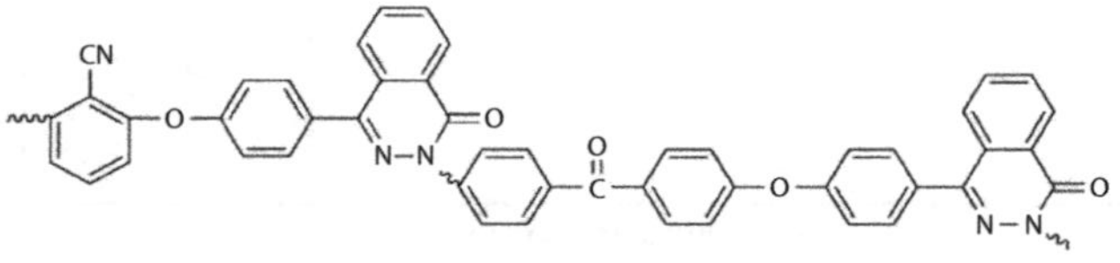	277	N/A	18 (flat sheet)	Superior mechanical strength and chemical resistance. Enhanced hydrophilicity due to presence of cyano group of strong polarity. Hu et al. (2012b)

(*Continued*)

TABLE 3.5 (*Continued*)

Overview of the Properties of Polymeric Substrate Used for TFC Membrane Preparation

Polymer[a]	Structure Formula	T_g (°C)[b]	MW (g mol^{-1})[c]	Polymer (wt%) in Dope Solution (Configuration)	Characteristics
PMDA/ODA PI (self-synthesized)		400	N/A	15 (flat sheet)	Very high T_g and excellent to most non-oxidizing acid and almost all organic solvents. Ba and Economy (2010)

[a] PAN, polyacrylonitrile; PEI, polyethylenimine; PP, polypropylene; PPSU, polyphenylsulfone; PPBES, copoly(phthalazinone biphenyl ether sulfone); PPESK, poly(phthalazinone ether sulfone ketone); PPENK, poly(phthalazinone ether nitrile ketone); PMDA/ODA PI, poly(pyromellitic dianhydride-co-4,4′-oxydianiline); PVDF, polyvinylidene difluoride.

[b] T_g—glass transition temperature.

[c] MW—molecular weight.

[d] The glass transition temperature of these polymers are obtained from Mulder (2003).

have also been conducted on substrate membrane with the aim of developing chemically and thermally stable substrates that are able to withstand harsh environments without compromising water permeability and rejection efficiency.

Generally, a thermally stable polymeric substrate is highly desired for many industrial processes. For instance, in the pulp and paper industry, a temperature of around 70–90°C is required in an integrated mill process. Most commercial membranes cannot withstand such a high temperature. The textile industry on the other hand discharges dyeing effluent at a temperature between 30°C and 80°C and pH in the range of 2–10. In 2010, Ba and Economy developed a poly(pyromellitic dianhydride-co-4,4′-oxydianiline) (PMDA/ODA PI)-based substrate from precursor polyamic acid in the presence of $ZnCl_2$ in the casting solution. Interaction between zinc ions and carboxylic groups of polyamic acid is found to be capable of creating an ionic CL structure, facilitating the formation of a supporting layer with improved surface properties. In comparison to the commercial PSF-based composite membrane, this thermally stable composite membrane provided a feasible way to significantly improve water permeability without any drop in salt rejection at elevated operating temperatures up to 95°C. The promising results can be attributed to the high mechanical strength of the substrate produced, which resists pore expansion at high temperature. Kim et al. (2009) on the other hand proposed to use modified polyvinylidene difluoride (PVDF) as an attractive candidate to replace PSF as a substrate because fluorinated polymers demonstrate relatively high mechanical strength and good resistance to chemicals compared to PSF-based substrate. Furthermore, PVDF can easily dissolve in common organic solvents; as a result, porous membranes can be produced via the phase inversion method by a simple immersion precipitation process. Since PVDF membrane is relatively hydrophobic, plasma treatment on the PVDF membrane surface is required prior to the IP process to improve its hydrophilicity so that it becomes suitable for TFC membrane preparation. Kim et al. (2009) reported that under 180 s plasma exposure using O_2/CH_4 gas mixture for PVDF substrate treatment, a TFC membrane made of this substrate could achieve outstanding performance, resulting in approximately 37% enhancement in pure water flux and slight increase in salt rejection compared to the membrane prepared using PSF-based substrate (0.19 L m^{-2} h^{-1} psi^{-1} and 96% @ 2000 ppm NaCl, 2.1 MPa).

Considering the active layer might be detached from the substrate under harsh conditions, Oh et al. (2001) modified the original functional group of polyacrylonitrile (PAN) substrate from –CN into –COOH by treating the substrate with NaOH solution at ambient temperature. With such treatment, the newly created functional groups could form a covalent or ionic bond with the PA layer, leading to improved adhesion extent and further demonstrated a better stability of performance during the filtration process. Li et al. (2015b) on the other hand modified the PES substrate surface with alkaline dopamine solution in order to render the substrate with reactive groups and improve its hydrophilicity. PD is claimed to be able to stick firmly onto PES substrate and its quinone groups could react with amino groups of PIP via the Michael addition, allowing covalent immobilization of the PIP monomer to the substrate during the IP process. As tested by alcohol treatment for 15 days, the TFC membrane made of modified PES substrate exhibited good structural stability with less than 5% decline in Na_2SO_4 rejection. The control TFC membrane meanwhile exhibited more than 50% decline in salt rejection. This promising data are mainly due to the strong interactions between the PA active layer and PD-modified PES substrate.

Using a novel synthesized poly(phthalazinone ether amide) (PPEA) as substrate material, a thermally stable composite membrane was successfully fabricated by Wu et al. (2009). The resultant composite membrane has shown superior performance in removing dyes from a dye–salt mixed solution at 1.0 MPa, 80°C. The relative stability of the flux and dye rejection during a 5-h experiment at 80°C testified to the thermal stability of the PPEA-based composite membrane. A series of new polymer materials produced by Dalian New Polymer Material Co. Ltd. (China) have also been examined as possible materials to fabricate a thermal-resistant substrate. Among these new materials, a copolymer—poly(phthalazinone ether sulfone ketone) (PPESK) was reported as exhibiting a higher upper temperature limit and better thermal stability than that of the PSF-based substrate (Wei et al., 2005). Even when the PPESK-based substrate was operated at a temperature greater than 80°C, there was no significant sign of pore expansion, proving its extraordinary resistance against thermal attack. Another thermally stable composite membrane prepared from poly(phthalazinone ether nitrile ketone) (PPENK) substrate also showed excellent thermal stability (Hu et al., 2012b). When tested at 1 MPa and with 1000 ppm Na_2SO_4 solution as feed, the salt rejection was kept at about 95% while the flux of the membrane increased significantly from 49.5 to 196.7 L m^{-2} h^{-1} with increasing temperature from 20°C to 80°C. The constant salt rejection of this work can be attributed to the improved binding between the substrate and the active layer. Higher water flux resulted from higher operating temperature can be primarily due to the higher diffusion rate of water molecules through the membrane. Similar findings on thermal stability were also reported by Han et al. (2011) when copoly(phthalazinone biphenyl ether sulfone) (PPBES) was used as support for composite NF membrane fabrication. The experimental results revealed that the flux of the membrane remarkably increased from 61 to 290 L m^{-2} h^{-1} with little reduction in Na_2SO_4 rejection when the operating temperature was increased from 18°C to 85°C at 1 MPa. The performance stability of the PPBES-based composite membrane was further compared with the PSF-based composite membrane at a temperature of 85°C. As expected, the flux of the PPBES-based composite membrane was kept relatively stable throughout the 72-h studied period while the PSF-based composite membrane was broken up, leading to significant flux increase and decrease in salt rejection.

It is also very attractive to apply polypropylene (PP) as a substrate for TFC membrane preparation, owing to its high durability and resistance to chemical, pH, and a wide range of solvents. However, the hydrophobic characteristic of PP membrane has limited its potential application in the preparation of composite membrane by IP. To overcome the problem, Korikov et al. (2006) modified the wettability of the PP surface through a chemical surface oxidation method in an effort to improve the adhesion between the PA selective layer and supporting layer. Kosaraju and Sirkar (2008) also utilized the PP membrane as a supporting substrate in making composite membranes of flat film and hollow fiber. The composite membranes prepared proved to be stable in methanol permeation and brilliant blue rejection even after 10-weeks exposure to toluene solution, showing the great potential of PP-based composite membranes in handling a solution that consists of solvent. In 2012, Pan et al. introduced peroxide onto the surface of PP fiber (to hydrophilize PP) using ozone treatment followed by grafting acrylamide. The results showed that the composite NF membrane prepared from this modified PP fiber support could reject around 65% Na_2SO_4 (2000 ppm) and 93% Fast Green FCF (MW: 808.84 g mol^{-1}, 10 ppm) at an operating pressure of 0.5 MPa. Other than the polymers mentioned above, alternative polymeric substrates reported with enhanced properties for TFC membrane fabrication include sulfonated poly(phthalazinone ether sulfone ketone) (SPPESK) (Dai et al., 2002), poly(tetrafluoroethylene) (PTFE) (Liu et al., 2008), PEI (Han, 2013), PI (Jimenez-Solomon et al., 2012; Hermans et al., 2015), polyphenylsulfone (PPSU) (Liu et al., 2012b), PES/polyaniline (PANI) blend (Zhu et al., 2015), cross-linked PI, and poly(ether ether ketone) (PEEK) (Jimenez-Solomon et al., 2013). It must be noted that substrates made of polymers such as PI and PEEK are particularly suitable for filtrations in organic solutions as they are stable in strong swelling solvents, namely hexane and heptane.

Recent developments in nanofibers production technology have opened up the possibilities of applying nanofibers for various process improvements, including as a support for TFC membrane. Compared to the substrate made via the phase inversion process, the nanofiber substrate that is electrospun exhibits amazing characteristics such as very large surface area to volume ratio, superior porosity, and excellent mechanical properties. In 2011, Bui et al. introduced a novel flat-sheet PA TFC membrane supported by a nonwoven web of electrospun nanofibers. The fibers were electrospun onto nonwoven polyester fabric followed by a PA layer coating onto the nanofiber support. Two polymeric materials (PSF and PES) were selected to prepare the nanofiber substrate and the results showed the PSF nanofibers demonstrated strong adhesion with the PA layer while PA layer delamination occurred on PES nanofibers. It is explained that the presence of BPA moiety in the PSF structure is the main factor contributing to better adhesion between PSF and PA derived from MPD and TMC through a specific chemical interaction as illustrated in Figure 3.4a. Figure 3.4b and c meanwhile shows the cross section and top surface of TFC membrane made of

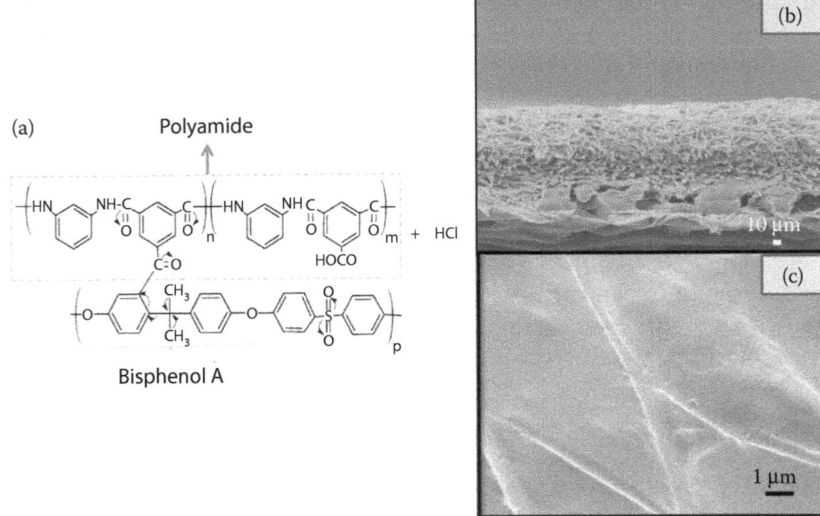

FIGURE 3.4
(a) Diagram of a possible CL interaction between PA and the BPA group of PSF (arrows show the proposed reaction mechanism), (b) and (c) scanning electron microscopy (SEM) images of cross section and top surface of PSF-based TFC membrane. (Reprinted from *Journal of Membrane Science*, 385–386, Bui, N.-N. et al., Electrospun nanofiber supported thin film composite membranes for engineered osmosis, 10–19, Copyright 2011, with permission from Elsevier.)

PSF nanofiber substrate. With respect to performance, the best nanofiber supported-PA composite membranes exhibited two to five times higher flux than a standard commercial membrane.

Wang et al. (2014) compared the performance of composite membrane made of multilayered nanofiber structure with the control membrane made of PAN microporous substrate. The newly fabricated membrane that consisted of a three-layered structure, including the nonwoven support, PAN electrospun mid-layer, and composite barrier layer based on cellulose nanofiber (CN) and PA matrix showed significantly higher flux than that of the membrane prepared by the conventional phase inversion substrate, possibly because of the water-channel structure induced at the interface between CN and PA which increased the permeability of the barrier layer. Nanofiber made of nylon 6,6 was also used to prepare a substrate owing to its intrinsic hydrophilicity and superior strength compared to other nanofiber materials. Huang and McCutcheon (2014) reported that besides exhibiting lower water contact angle (around 38°), the tensile strength of nylon 6,6 nanofiber (10 MPa) was found to be much higher that those of nanofibers made of PSF (1.8 MPa) and PAN (5.7 MPa), making it highly desirable as TFC membrane support. When tested in forward osmosis (FO) process, the best TFC membrane made of nylon 6,6 nanofiber outperformed the standard commercial membrane by exhibiting a 1.5–2 fold enhanced water flux and an equal or lower specific salt flux. It is believed that such performance and tolerance to low selectivity could make this membrane an excellent candidate for further exploration in the NF process.

As substrate membrane can be prepared from various polymer materials and in different formats (e.g., flat sheet, hollow fiber, and nanofiber), the selection of a substrate for TFC membrane preparation is, therefore, dependent on the area of industrial application as well as the manufacturing and material cost. Nevertheless, advances in the development of stable substrate will offer opportunities to expand the application area of membranes particularly to industries where there are harsh conditions, for example, the petrochemical and pharmaceutical industries.

3.3.2 Polymer/Inorganic Nanocomposite Substrate

This class of nanocomposite substrate was first developed by Pendergast et al. (2010) to study the effects of nanomaterials on the membrane compaction behavior during the RO process. In this study, silica or zeolite nanoparticles were embedded into the PSF substrate, which was then used in the IP process to prepare composite membranes. Besides exhibiting higher water flux, the prepared membranes experienced less flux decline during the compaction when compared with the original TFC membrane. The existence of nanofillers was believed to have provided necessary mechanical support to mitigate the collapse of the porous structure and thickness reduction upon high-pressure operation. Electron microscope images meanwhile supported the hypothesis that nanocomposite supports resisted physical compaction.

Compared with TFC NF or RO membranes, the effect of nanomaterials on the substrate properties of FO membrane is much more significant as both the top PA selective layer and bottom substrate are simultaneously contacted with aqueous solutions during the filtration process, but of different osmotic pressures. Developing a substrate with enhanced hydrophilicity and reduced structural parameter (S) (which is controlled by thickness, tortuosity, and porosity) is, therefore, important to mitigate internal concentration polarization during the filtration process that occurs inside the substrate membrane (Yin and Deng, 2015).

Similar to PA nanocomposite layer development which aims to improve composite membrane separation properties, substrate modification using inorganic nanofillers does alter the chemistry of the PA layer formed over the nanocomposite substrate to a certain extent, leading to the improved performance of the pressure-driven membrane process. Kim et al. (2012) and Son et al. (2015) have incorporated MWCNTs into microporous substrate made of PSF and PES, respectively, aiming to develop an enhanced composite membrane performance for water purification and separation. In comparison to the control TFC membrane, the newly developed composite membrane displayed higher water flux without sacrificing salt rejection, owing to the improved surface hydrophilicity, average pore width, and porosity in the MWCNTs-modified substrate (Son et al., 2015). Higher negative surface charge of the PA layer was also reported in that work, but in-depth analysis was not performed to explain how the MWCNTs-modified substrate with increased charge properties could render the coated PA layer with thickness of several hundreds of nanometers to possess a greater negative charge.

With respect to the nanofiller loadings used in nanocomposite substrate preparation, Emadzadeh et al. (2014) found that 0.6 wt% TiO_2 nanoparticles is the optimum loading to produce a composite membrane with a good balance of water permeability and salt rejection. When that amount of TiO_2 was introduced into the PSF substrate, the water permeability of new composite membrane was significantly increased to 6.57 L m^{-2} h^{-1}, that is, 100% higher than that of the control TFC membrane (3.27 L m^{-2} h^{-1}) when tested using 20 mM NaCl feed solution at 2.5 bar. This flux enhancement can be explained by the improved hydrophilicity of the substrate together with the increased overall porosity upon TiO_2 incorporation. Excessive use of TiO_2 nanoparticles (0.9 wt%) is not recommended as it tends to cause significant nanoparticle agglomeration on the substrate surface, leading to increase in surface roughness. This, as a result, negatively affects the degree of PA CL and further reduces the salt removal rate. The impact of adding zeolite NaY nanoparticles into the substrate matrix on RO membrane performance can also be found elsewhere (Ma et al., 2013), where it was reported that 0.5 wt% was the ideal nanofiller loading for nanocomposite substrate in order to achieve a good combination of water flux and solute rejection in TFN membrane.

An attempt was also made to incorporate GO into the PSF substrate for the preparation of a new type of NF membrane (Lai et al., 2016). The results revealed that the presence of a small quantity of GO (0.3 wt%) in the PSF substrate was able to improve the physicochemical properties of the PSF substrate and PA selective layer with respect to structural integrity (of the microporous substrate), surface hydrophilicity, roughness, and charge. When benchmarking with the control TFC membrane, the pure water flux of the best-performing GO-embedded TFC membrane was reported to improve by 50.9% without compromising the salt rejection. This GO-embedded TFC membrane also showed high rejection toward multivalent salts, that is, 95.2% and 91.1% for Na_2SO_4 and $MgSO_4$, respectively. It was also capable of rejecting close to 60% of NaCl, that is, 88.5% higher than that of the control TFC membrane. The promising outcome from the filtration experiments can be mainly attributed to the unique characteristics of GO nanosheets such as superior hydrophilicity and high negative charge. The charge value of GO (–30 to –58 mV at pH 5–8) is generally much higher than those of TiO_2 nanoparticles (30 to –15 mV at pH 5–7) and CNTs (–8.3 mV at pH 7).

As a chapter conclusion, it is safe to say that NF technology is a mature separation and purification process that can be implemented in different industrial applications without major issues regarding operation and permeate quality, but there is still plenty of room for further improvement on membrane properties using advanced materials. Continuous improvement in the intrinsic characteristics of NF membrane as well as its separation properties is still relevant in order to widen its application to more challenging areas.

References

Aburabie, J., Neelakanda, P., Karunakaran, M., Peinemann, K.-V. 2015. Thin-film composite cross-linked polythiosemicarbazide membranes for organic solvent nanofiltration (OSN). *Reactive and Functional Polymers*. 86, 225–232.

Abu Seman, M.N., Khayet, M., Hilal, N. 2010. Nanofiltration thin-film composite polyester polyethersulfone-based membranes prepared by interfacial polymerization. *Journal of Membrane Science*. 348, 109–116.

Abu Seman, M.N., Khayet, M., Hilal, N. 2011. Development of antifouling properties and performance of nanofiltration membranes modified by interfacial polymerisation. *Desalination*. 273, 36–47.

Abu Tarboush, B.J., Rana, D., Matsuura, T., Arafat, H.A., Narbaitz, R.M. 2008. Preparation of thin film composite polyamide membranes for desalination using novel hydrophilic surface modifying marcomolecules. *Journal of Membrane Science*. 325, 166–175.

Ba, C., Economy, J. 2010. Preparation of PMDA/ODA polyimide membrane for use as substrate in a thermally stable composite reverse osmosis membrane. *Journal of Membrane Science*. 363, 140–148.

Baroña, G.N.B., Choi, M., Jung, B. 2012. High permeate flux of PVA/PSf thin film composite nanofiltration membrane with aluminosilicate single-walled nanotubes. *Journal of Colloid and Interface Science*. 386, 189–197.

Basu, S., Maes, M., Cano-Odena, A., Alaerts, L., De Vos, D.E., Vankelecom, I.F.J. 2009. Solvent resistant nanofiltration (SRNF) membranes based on metal-organic frameworks. *Journal of Membrane Science*. 344, 190–198.

Black, L.E. 1991. Interfacially polymerized membranes for reverse osmosis separation of organic solvent solutions. US Patent No. 5173191.

Buch, P.R., Jagan Mohan, D., Reddy, A.V.R. 2008. Preparation, characterization and chlorine stability of aromatic–cycloaliphatic polyamide thin film composite membranes. *Journal of Membrane Science*. 309, 36–44.

Bui, N.-N., Lind, M.L., Hoek, E.M.V., McCutcheon, J.R. 2011. Electrospun nanofiber supported thin film composite membranes for engineered osmosis. *Journal of Membrane Science*. 385–386, 10–19.

Chen, G., Li, S., Zhang, X., Zhang, S. 2008. Novel thin-film composite membranes with improved water flux from sulfonated cardo poly(arylene ether sulfone) bearing pendant amino groups. *Journal of Membrane Science*. 310, 102–109.

Chen, X.Q., Liu, Y., Guo, Z., Shao, L. 2015. Nanofiltration membrane achieving dual resistance to fouling and chlorine for "green" separation of antibiotics. *Journal of Membrane Science*. 493, 156–166.

Dai, Y., Jian, X., Zhang, S., Guiver, M.D. 2002. Thin film composite (TFC) membranes with improved thermal stability from sulfonated poly(phthalazinone ether sulfone ketone) (SPPESK). *Journal of Membrane Science*. 207, 189–197.

Dalwani, M., Benes, N.E., Bargeman, G., Stamatialis, D., Wessling, M. 2011. Effect of pH on the performance of polyamide/polyacrylonitrile based thin film composite membranes. *Journal of Membrane Science*. 372, 228–238.

Duan, M., Wang, Z., Xu, J., Wang, J., Wang, S. 2010. Influence of hexamethyl phosphoramide on polyamide composite reverse osmosis membrane performance. *Separation and Purification Technology*. 75, 145–155.

Emadzadeh, D., Lau, W.J., Matsuura, T., Ismail, A.F., Rahbari-Sisakht, M. 2014. Synthesis and characterization of thin film nanocomposite forward osmosis membrane with hydrophilic nanocomposite support to reduce internal concentration polarization. *Journal of Membrane Science*. 449, 74–85.

Fathizadeh, M., Aroujalian, A., Raisi, A. 2011. Effect of added NaX nano-zeolite into polyamide as a top thin layer of membrane on water flux and salt rejection in a reverse osmosis process. *Journal of Membrane Science*. 375, 88–95.

Geens, J., De Witte, B., Van der Bruggen, B. 2007. Removal of API's (active pharmaceutical ingredients) from organic solvents by nanofiltration. *Separation Science and Technology*. 42, 2435–2449.

Ghosh, A.K., Hoek, E.M.V. 2009. Impacts of support membrane structure and chemistry on polyamide–polysulfone interfacial composite membranes. *Journal of Membrane Science*. 336, 140–148.

Han, R. 2013. Formation and characterization of (melamine–TMC) based thin film composite NF membranes for improved thermal and chlorine resistances. *Journal of Membrane Science*. 425–426, 176–181.

Han, R., Zhang, S., Hu, L., Guan, S., Jian, X. 2011. Preparation and characterization of thermally stable poly(piperazine amide)/PPBES composite nanofiltration membrane. *Journal of Membrane Science*. 370, 91–96.

Hermans, S., Dom, E., Mariën, H., Koeckelberghs, G., Vankelecom, I.F.J. 2015. Efficient synthesis of interfacially polymerized membranes for solvent resistant nanofiltration. *Journal of Membrane Science*. 476, 356–363.

Ho, S., Kwak, S., Sohn, B., Hyun, T. 2003. Design of TiO_2 nanoparticle self-assembled aromatic polyamide thin-film-composite (TFC) membrane as an approach to solve biofouling problem. *Journal of Membrane Science*. 211, 157–165.

Hu, D., Xu, Z.-L., Chen, C. 2012a. Polypiperazine-amide nanofiltration membrane containing silica nanoparticles prepared by interfacial polymerization. *Desalination*. 301, 75–81.

Hu, L., Zhang, S., Han, R., Jian, X. 2012b. Preparation and performance of novel thermally stable polyamide/PPENK composite nanofiltration membranes. *Applied Surface Science*. 258, 9047–9053.

Huang, L., McCutcheon, J.R. 2014. Hydrophilic nylon 6,6 nanofibers supported thin film composite membranes for engineered osmosis. *Journal of Membrane Science*. 457, 162–169.

Jadav, G.L., Singh, P.S. 2009. Synthesis of novel silica–polyamide nanocomposite membrane with enhanced properties. *Journal of Membrane Science*. 328, 257–267.

Jegal, J., Min, S.G., Lee, K.-H. 2002. Factors affecting the interfacial polymerization of polyamide active layers for the formation of polyamide composite membranes. *Journal of Applied Polymer Science*. 86, 2781–2787.

Jeong, B.-H., Hoek, Eric M.V., Yan, Y., Subramani, A., Huang, X., Hurwitz, G., Ghosh, A.K., Jawor, A. 2007. Interfacial polymerization of thin film nanocomposites: A new concept for reverse osmosis membranes. *Journal of Membrane Science*. 294, 1–7.

Jimenez-Solomon, M.F., Bhole, Y., Livingston, A.G. 2012. High flux membranes for organic solvent nanofiltration (OSN)—Interfacial polymerization with solvent activation. *Journal of Membrane Science*. 423–424, 371–382.

Jimenez-Solomon, M.F., Gorgojo, P., Munoz-Ibanez, M., Livingston, A.G. 2013. Beneath the surface: Influence of supports on thin film composite membranes by interfacial polymerization for organic solvent nanofiltration. *Journal of Membrane Science*. 448, 102–113.

Jin, J., Liu, D., Zhang, D., Yin, Y., Zhao, X., Zhang, Y. 2015. Preparation of thin-film composite nanofiltration membranes with improved antifouling property and flux using 2,2′-oxybis-ethylamine. *Desalination*. 355, 141–146.

Kim, E.-S., Deng, B. 2011. Fabrication of polyamide thin-film nano-composite (PA-TFN) membrane with hydrophilized ordered mesoporous carbon (H-OMC) for water purifications. *Journal of Membrane Science*. 375, 46–54.

Kim, E.-S., Hwang, G., El-Din, M.G., Liu, Y. 2012. Development of nanosilver and multi-walled carbon nanotubes thin-film nanocomposite membrane for enhanced water treatment. *Journal of Membrane Science*. 394–395, 37–48.

Kim, E.-S., Kim, Y.J., Yu, Q., Deng, B. 2009. Preparation and characterization of polyamide thin-film composite (TFC) membranes on plasma-modified polyvinylidene fluoride (PVDF). *Journal of Membrane Science*. 344, 71–81.

Kim, I.-C., Jeong, B.-R., Kim, S.-J., Lee, K.-H. 2013. Preparation of high flux thin film composite polyamide membrane: The effect of alkyl phosphate additives during interfacial polymerization. *Desalination*. 308, 111–114.

Kim, S.H.O. 2005. Positron annihilation spectroscopic evidence to demonstrate the flux-enhancement mechanism in membrane. *Environmental Science and Technology*. 39, 1764–1770.

Kong, C., Kanezashi, M., Yamomoto, T., Shintani, T., Tsuru, T. 2010. Controlled synthesis of high performance polyamide membrane with thin dense layer for water desalination. *Journal of Membrane Science*. 362, 76–80.

Kong, C., Shintani, T., Kamada, T., Freger, V., Tsuru, T. 2011. Co-solvent-mediated synthesis of thin polyamide membranes. *Journal of Membrane Science*. 384, 10–16.

Korikov, A.P., Kosaraju, P.B., Sirkar, K.K. 2006. Interfacially polymerized hydrophilic microporous thin film composite membranes on porous polypropylene hollow fibers and flat films. *Journal of Membrane Science*. 279, 588–600.

Kosaraju, P.B., Sirkar, K.K. 2008. Interfacially polymerized thin film composite membranes on microporous polypropylene supports for solvent-resistant nanofiltration. *Journal of Membrane Science*. 321, 155–161.

La, Y.-H., Sooriyakumaran, R., Miller, D.C., Fujiwara, M., Terui, Y., Yamanaka, K., McCloskey, B.D., Freeman, B.D., Allen, R.D. 2010. Novel thin film composite membrane containing ionizable hydrophobes: pH-dependent reverse osmosis behavior and improved chlorine resistance. *Journal of Materials Chemistry*. 20, 4615–4620.

Lai, G.S., Lau, W.J., Goh, P.S., Ismail, A.F., Yusof, N., Tan, Y.H. 2016. Graphene oxide incorporated thin film nanocomposite nanofiltration membrane for enhanced salt removal performance. *Desalination*. 387, 14–24.

Lee, H.S., Im, S.J., Kim, J.H., Kim, H.J., Kim, J.P., Min, B.R. 2008. Polyamide thin-film nanofiltration membranes containing TiO_2 nanoparticles. *Desalination*. 219, 48–56.

Lee, K.P., Zheng, J., Bargeman, G., Kemperman, A.J.B., Benes, N.E. 2015. pH stable thin film composite polyamine nanofiltration membranes by interfacial polymerization. *Journal of Membrane Science*. 478, 75–84.

Lee, S.Y., Kim, H.J., Patel, R., Im, S.J., Kim, J.H., Min, B.R. 2007. Silver nanoparticles immobilized on thin film composite polyamide membrane: Characterization, nanofiltration, antifouling properties. *Polymers for Advanced Technologies*. 18, 562–568.

Li, L., Zhang, S., Zhang, X. 2009. Preparation and characterization of poly(piperazine-amide) composite nanofiltration membrane by interfacial polymerization of 3,3′,5,5′-biphenyl tetraacyl chloride and piperazine. *Journal of Membrane Science*. 335, 133–139.

Li, L., Zhang, S., Zhang, X., Zheng, G. 2008. Polyamide thin film composite membranes prepared from isomeric biphenyl tetraacyl chloride and m-phenylene-diamine. *Journal of Membrane Science*. 315, 20–27.

Li, X., Zhang, C., Zhang, S., Li, J., He, B., Cui, Z. 2015a. Preparation and characterization of positively charged polyamide composite nanofiltration hollow fiber membrane for lithium and magnesium separation. *Desalination*. 369, 26–36.

Li, Y., Su, Y., Dong, Y., Zhao, X., Jiang, Z., Zhang, R., Zhao, J. 2014. Separation performance of thin-film composite nanofiltration membrane through interfacial polymerization using different amine monomers. *Desalination*. 333, 59–65.

Li, Y., Su, Y., Li, J., Zhao, X., Zhang, R., Fan, X., Zhu, J., Ma, Y., Liu, Y., Jiang, Z. 2015b. Preparation of thin film composite nanofiltration membrane with improved structural stability through the mediation of polydopamine. *Journal of Membrane Science*. 476, 10–19.

Liu, M., Wu, D., Yu, S., Gao, C. 2009. Influence of the polyacyl chloride structure on the reverse osmosis performance, surface properties and chlorine stability of the thin-film composite polyamide membranes. *Journal of Membrane Science*. 326, 205–214.

Liu, M., Yao, G., Cheng, Q., Ma, M., Yu, S., Gao, C. 2012a. Acid stable thin-film composite membrane for nanofiltration prepared from naphthalene-1,3,6-trisulfonylchloride (NTSC) and piperazine (PIP). *Journal of Membrane Science*. 415–416, 122–131.

Liu, M., Zheng, Y., Shuai, S., Zhou, Q., Yu, S., Gao, C. 2012c. Thin-film composite membrane formed by interfacial polymerization of polyvinylamine (PVAm) and trimesoyl chloride (TMC) for nanofiltration. *Desalination*. 288, 98–107.

Liu, T.-Y., Bian, L.-X., Yuan, H.-G., Pang, B., Lin, Y.-K., Tong, Y., Van der Bruggen, B., Wang, X.-L. 2015. Fabrication of a high-flux thin film composite hollow fiber nanofiltration membrane for wastewater treatment. *Journal of Membrane Science*. 478, 25–36.

Liu, Y., Zhang, S., Zhou, Z., Ren, J., Geng, Z., Luan, J., Wang, G. 2012b. Novel sulfonated thin-film composite nanofiltration membranes with improved water flux for treatment of dye solutions. *Journal of Membrane Science*. 394–395, 218–229.

Liu, Y.-L., Yu, C.-H., Lai, J.-Y. 2008. Poly(tetrafluoroethylene)/polyamide thin-film composite membranes via interfacial polymerization for pervaporation dehydration on an isopropanol aqueous solution. *Journal of Membrane Science*. 315, 106–115.

Ma, N., Wei, J., Qi, S., Zhao, Y., Gao, Y., Tang, C.Y. 2013. Nanocomposite substrates for controlling internal concentration polarization in forward osmosis membranes. *Journal of Membrane Science*. 441, 54–62.

Mansourpanah, Y., Alizadeh, K., Madaeni, S.S., Rahimpour, A., Soltani Afarani, H. 2011. Using different surfactants for changing the properties of poly(piperazineamide) TFC nanofiltration membranes. *Desalination*. 271, 169–177.

Mansourpanah, Y., Madaeni, S.S., Rahimpour, A. 2009. Fabrication and development of interfacial polymerized thin-film composite nanofiltration membrane using different surfactants in organic phase; study of morphology and performance. *Journal of Membrane Science*. 343, 219–228.

Maurya, S.K., Parashuram, K., Singh, P.S., Ray, P., Reddy, A.V.R. 2012. Preparation of polysulfone–polyamide thin film composite hollow fiber nanofiltration membranes and their performance in the treatment of aqueous dye solutions. *Desalination*. 304, 11–19.

Misdan, N., Lau, W.J., Ismail, A.F., Matsuura, T., Rana, D. 2014. Study on the thin film composite poly(piperazine-amide) nanofiltration membrane: Impacts of physicochemical properties of substrate on interfacial polymerization formation. *Desalination*. 344, 198–205.

Mulder, M. 2003. *Basic Principles of Membrane Technology*. Kluwer Academic Publishers, Dordrecht, The Netherlands.

Namvar-Mahboub, M., Pakizeh, M., Davari, S. 2014. Preparation and characterization of UZM-5/polyamide thin film nanocomposite membrane for dewaxing solvent recovery. *Journal of Membrane Science*. 459, 22–32.

Oh, N.-W., Jegal, J., Lee, K.-H. 2001. Preparation and characterization of nanofiltration composite membranes using polyacrylonitrile (PAN). II. Preparation and characterization of polyamide composite membranes. *Journal of Applied Polymer Science*. 80, 2729–2736.

Ong, C.S., Lau, W.J., Ismail, A.F. 2012. Treatment of dyeing solution by NF membrane for decolorization and salt reduction. *Desalination and Water Treatment*. 50, 245–253.

Pendergast, M.T.M., Nygaard, J.M., Ghosh, A.K., Hoek, E.M.V. 2010. Using nanocomposite materials technology to understand and control reverse osmosis membrane compaction. *Desalination*. 261, 255–263.

Peng, J.M., Su, Y.L., Chen, W.J., Zhao, X.T., Jiang, Z.Y., Dong, Y.N., Zhang, Y., Liu, J.Z., Cao, X.Z. 2013. Polyamide nanofiltration membrane with high separation performance prepared by EDC/NHS mediated interfacial polymerization. *Journal of Membrane Science*. 427, 92–100.

Petersen, R.J. 1993. Composite reverse osmosis and nanofiltration membranes. *Journal of Membrane Science*. 83, 81–150.

Peyravi, M., Jahanshahi, M., Rahimpour, A., Javadi, A., Hajavi, S. 2014. Novel thin film nanocomposite membranes incorporated with functionalized TiO_2 nanoparticles for organic solvent nanofiltration. *Chemical Engineering Journal*. 241, 155–166.

Peyravi, M., Rahimpour, A., Jahanshahi, M. 2012. Thin film composite membranes with modified polysulfone supports for organic solvent nanofiltration. *Journal of Membrane Science*. 423–424, 225–237.

Pourjafar, S., Rahimpour, A., Jahanshahi, M. 2012. Synthesis and characterization of PVA/PES thin film composite nanofiltration membrane modified with TiO_2 nanoparticles for better performance and surface properties. *Journal of Industrial and Engineering Chemistry*. 18, 1398–1405.

Rajaeian, B., Rahimpour, A., Tade, M.O., Liu, S. 2013. Fabrication and characterization of polyamide thin film nanocomposite (TFN) nanofiltration membrane impregnated with TiO_2 nanoparticles. *Desalination*. 313, 176–188.

Rana, D., Kim, Y., Matsuura, T., Arafat, H.A. 2011. Development of antifouling thin-film-composite membranes for seawater desalination. *Journal of Membrane Science*. 367, 110–118.

Roy, S., Ntim, S.A., Mitra, S., Sirkar, K.K. 2011. Facile fabrication of superior nanofiltration membranes from interfacially polymerized CNT–polymer composites. *Journal of Membrane Science*. 375, 81–87.

Saha, N.K., Joshi, S.V. 2009. Performance evaluation of thin film composite polyamide nanofiltration membrane with variation in monomer type. *Journal of Membrane Science*. 342, 60–69.

Sani, N.A.A., Lau, W.J., Ismail, A.F. 2015. Polyphenylsulfone-based solvent resistant nanofiltration (SRNF) membrane incorporated with copper-1,3,5-benzenetricarboxylate (Cu-BTC) nanoparticles for methanol separation. *RSC Advances*. 5, 13000–13010.

Son, M., Choi, H., Liu, L., Celik, E., Park, H., Choi, H. 2015. Efficacy of carbon nanotube positioning in the polyethersulfone support layer on the performance of thin-film composite membrane for desalination. *Chemical Engineering Journal*. 266, 376–384.

Sun, S.P., Hatton, T.A., Chan, S.Y., Chung, T.-S. 2012. Novel thin-film composite nanofiltration hollow fiber membranes with double repulsion for effective removal of emerging organic matters from water. *Journal of Membrane Science*. 401–402, 152–162.

Tang, B., Huo, Z., Wu, P. 2008. Study on a novel polyester composite nanofiltration membrane by interfacial polymerization of triethanolamine (TEOA) and trimesoyl chloride (TMC). *Journal of Membrane Science*. 320, 198–205.

Tang, B., Zou, C., Wu, P. 2010. Study on a novel polyester composite nanofiltration membrane by interfacial polymerization. II. The role of lithium bromide in the performance and formation of composite membrane. *Journal of Membrane Science*. 365, 276–285.

Wang, H., Li, L., Zhang, X., Zhang, S. 2010. Polyamide thin-film composite membranes prepared from a novel triamine 3,5-diamino-N-(4-aminophenyl)-benzamide monomer and m-phenylenediamine. *Journal of Membrane Science*. 353, 78–84.

Wang, H., Zhang, Q., Zhang, S. 2011. Positively charged nanofiltration membrane formed by interfacial polymerization of 3,3′,5,5′-biphenyl tetraacyl chloride and piperazine on a poly(acrylonitrile) (PAN) support. *Journal of Membrane Science*. 378, 243–249.

Wang, X., Fang, D., Hsiao, B.S., Chu, B. 2014. Nanofiltration membranes based on thin-film nanofibrous composites. *Journal of Membrane Science*. 469, 188–197.

Wei, J., Jian, X., Wu, C., Zhang, S., Yan, C. 2005. Influence of polymer structure on thermal stability of composite membranes. *Journal of Membrane Science*. 256, 116–121.

Wu, C., Zhang, S., Yang, D., Jian, X. 2009. Preparation, characterization and application of a novel thermal stable composite nanofiltration membrane. *Journal of Membrane Science*. 326, 429–434.

Wu, H., Tang, B., Wu, P. 2010. MWNTs/polyester thin film nanocomposite membrane: An approach to overcome the trade-off effect between permeability and selectivity. *The Journal of Physical Chemistry C*. 114, 16395–16400.

Wu, H., Tang, B., Wu, P. 2013. Optimizing polyamide thin film composite membrane covalently bonded with modified mesoporous silica nanoparticles. *Journal of Membrane Science*. 428, 341–348.

Xia, S., Yao, L., Zhao, Y., Li, N., Zheng, Y. 2015. Preparation of graphene oxide modified polyamide thin film composite membranes with improved hydrophilicity for natural organic matter removal. *Chemical Engineering Journal*. 280, 720–727.

Xiang, J., Xie, Z., Hoang, M., Ng, D., Zhang, K. 2014. Effect of ammonium salts on the properties of poly(piperazzineamide) thin film composite nanofiltration membrane. *Journal of Membrane Science*. 465, 34–40.

Yang, F., Zhang, S., Yang, D., Jian, X. 2007. Preparation and characterization of polypiperazine amide/PPESK hollow fiber composite nanofiltration membrane. *Journal of Membrane Science*. 301, 85–92.

Yin, J., Deng, B. 2015. Polymer-matrix nanocomposite membranes for water treatment. *Journal of Membrane Science*. 479, 256–275.

Yu, S., Liu, M., Lü, Z., Zhou, Y., Gao, C. 2009. Aromatic–cycloaliphatic polyamide thin-film composite membrane with improved chlorine resistance prepared from m-phenylenediamine-4-methyl and cyclohexane-1,3,5-tricarbonyl chloride. *Journal of Membrane Science*. 344, 155–164.

Yu, S., Ma, M., Liu, J., Tao, J., Liu, M., Gao, C. 2011. Study on polyamide thin-film composite nanofiltration membrane by interfacial polymerization of polyvinylamine (PVAm) and isophthaloyl chloride (IPC). *Journal of Membrane Science*. 379, 164–173.

Zhao, J., Su, Y., He, X., Zhao, X., Li, Y., Zhang, R., Jiang, Z. 2014. Dopamine composite nanofiltration membranes prepared by self-polymerization and interfacial polymerization. *Journal of Membrane Science*. 465, 41–48.

Zhou, C., Shi, Y., Sun, C., Yu, S., Liu, M. Gao, C. 2014. Thin-film composite membranes formed by interfacial polymerization with natural material sericin and trimesoyl chloride for nanofiltration. *Journal of Membrane Science*. 471, 381–391.

Zhu, S., Zhao, S., Wang, Z., Tian, X., Shi, M., Wang, J., Wang, S. 2015. Improved performance of polyamide thin-film composite nanofiltration membrane by using polyetersulfone/polyaniline membrane as the substrate. *Journal of Membrane Science*. 493, 263–274.

4

Technical Challenges and Approaches in Nanofiltration Membrane Fabrication

4.1 TFC Hollow Fiber Membrane

4.1.1 Challenges

The commercially available TFC NF membranes are prepared by forming a very thin PA active layer on a porous support membrane via an IP technique. The technique offers outstanding advantages as either the top selective layer or bottom porous substrate of membrane can be independently modified and optimized to enhance the water permeation rate and solute rejection while offering excellent mechanical strength and/or compression resistance.

Unlike TFC membrane in plate and frame or spiral wound configurations, composite membrane in hollow fiber configuration is experiencing a slow growth in applications, primarily due to the differences in structural geometry and handling of solution flow during the IP process. Works have been conducted to produce composite flat membranes using the conventional IP technique, but if this technique is applied directly to the outer or lumen surface of a hollow fiber microporous membrane without any modification, the formation of a defect-free PA film is almost impossible. The formed PA film would not adhere properly to the hollow fiber substrate during the fabrication process as the rubber roller which is generally applied in the TFC flat sheet membrane could not be utilized to remove the excess amine-type monomer from the substrate surface. Without the use of the rubber roller, pinholes (caused by the droplets of amine aqueous solution) are likely to form on the fiber geometry. Such a "defect" in the thin active film would lead to a bulk flow of impurities and contaminate the permeated solution. Although filter paper has been proposed to remove excess aqueous solution from the support membrane, this method is only applicable to the outer surface of the hollow fiber and is not practical for the industrial manufacturing process.

Compared to flat sheet membranes, hollow fiber membranes exhibit a higher packing density, a higher surface area to volume ratio and self-support capability, and are more cost-effective in large-scale production and operation. Therefore, successful fabrication of TFC hollow fiber membranes via the IP technique could offer wider range of industrial applications.

4.1.2 Innovative Approaches

To address the challenges of establishing a PA layer either on the outer or lumen surface of hollow fiber membranes, the conventional IP procedure has to be modified in order to achieve the desirable PA layer formation.

One of the earliest mentions of the TFC hollow fiber membrane was disclosed in a US patent in 1990 in which the inventors developed a coating system that could form a PA thin layer on a hollow fiber outer surface in a continuous manner (Tadros and Trehu, 1990). The fabrication process is briefly described as follows. The hollow fiber membrane which acts as a support layer is first fixed in a bobbin. It is then guided through a container completely filled with an aqueous solution of monomer before passing through a secondary solution containing another monomer. Spontaneous drying is performed on the fiber surface soon after it passes the secondary solution. During the process, the gas phase is avoided between the two solutions as the excess aqueous solution on the fiber surface is removed by capillary action. Continuous hollow fiber coating process was also documented in another US patent filed by Kumano et al. (1998). Compared to the conventional IP process that uses two types of solutions, a third liquid which is substantially immiscible with the first and second solution is introduced to improve the PA characteristics on the surface of the hollow fiber membrane that is produced in a continuous and stable manner.

Although the potential applications of TFC hollow fiber membrane with a PA layer formed on its outer layer have been reported in many research articles, forming a defect-free PA layer onto a hollow fiber outer surface is much more challenging than forming a thin film onto its lumen surface. This is mainly due to the high risk of coating imperfection on the outer surface of a bundle of fibers that usually stick close to each other. Furthermore, in the case of forming a PA layer onto a fiber outer surface in a continuous manner, the thin film might be possibly peeled off and/or damaged when fiber outer surface gets in touch with driving rollers during the spinning process.

In comparison to the hollow fiber membrane with PA coated onto its outer surface, synthesizing TFC hollow fiber membrane with PA selective layer in the lumen surface seems to be more practical and worthy of study. Veríssimo et al. (2005), for example, modified the conventional IP process by introducing an intermediate organic solvent between the aqueous amine solution and the acid chloride solution, aiming to form a film on the inner surface of the support fiber without pinholes. The formation of PA film was confirmed by SEM and atomic force microscopy analyses. Yang et al. (2007) applied an easier method to remove the excess piperazine (PIP) solution and droplets by flushing nitrogen gas slowly through the lumen side of the fibers. The results showed that this modified IP procedure is able to produce composite hollow fiber membrane with better stability for long-term running, mainly due to the good compatibility between the PA active layer and substrate membrane. Li and Chung (2014) used compressed air to remove excess *m*-phenylenediamine (MPD) solution from the fiber lumen before introducing a hexane solution that contained 0.2 w/v% TMC. Compressed air was also used to sweep impurities on the inner surface of hollow fibers before the IP process started. Using this technique, it was reported that a PA layer thickness of around 100 nm could be produced. The approach of using compressed air to remove excess aqueous solution off the lumen side can also be found in other works (Shao et al., 2013; Sukitpaneenit and Chung, 2014).

From the viewpoint of large-scale industrial production, the current technology of composite hollow fiber membrane preparation is yet to reach a degree of maturity. The reproducibility might be the main manufacturing concern for this membrane to be produced. This area, however, is still one of the subjects deserving a better level of understanding in order to make full use of hollow fiber membrane prepared via a composite approach.

4.2 TFN Flat Sheet Membrane

4.2.1 Challenges

The potential of PA TFN membrane for water treatment process in particular has been demonstrated by many researchers since 2007, but there remain several key challenges related to TFN membrane making. Addressing these challenges is the key to further development of TFN membrane for industrial applications. This section will focus on the challenges of TFN membrane making in flat sheet format as TFN membrane in hollow fiber configuration is rarely found in the literature.

One of the main problems encountered during TFN membrane fabrication is the agglomeration of nanofillers in the PA layer. The TEM images in Figure 4.1 show particle agglomeration in the thin PA layer, which is likely to reduce the active surface area of nanoparticles and/or even result in the formation of defects (holes) in the PA structure. The main reason attributed to this phenomenon is the low dispersion rate of nanomaterials in the solutions (either aqueous or organic) used in the IP process. The nonuniform dispersion of particles, particularly in nonpolar organic solvent, makes them very easy to aggregate and unable to uniformly spread out on the top substrate surface. As a result, a certain part of the PA dense layer contains no nanomaterials at all.

It is generally known that hydrophilic nanomaterials can be better dispersed in the aqueous phase than in the organic phase, but when excessive aqueous phase is removed by applying a soft rubber roller on the

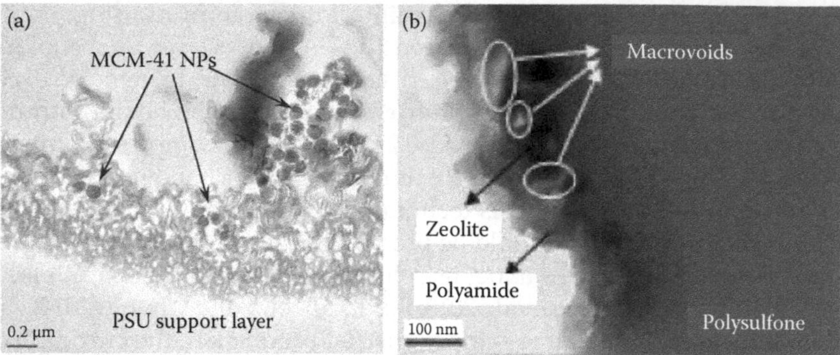

FIGURE 4.1
TEM images of the cross section of (a) TFN membrane with silica nanoparticles (MCM-41) embedded (Reprinted from *Journal of Membrane Science*, 423–424, Yin, J. et al., Fabrication of a novel thin-film nanocomposite (TFN) membrane containing MCM-41 silica nanoparticles (NPs) for water purification, 238–246, Copyright 2012, with permission from Elsevier.) and (b) TFN membrane with NaA zeolite embedded. (Huang, H. et al. 2013. Role of NaA zeolites in the interfacial polymerization process towards a polyamide nanocomposite reverse osmosis membrane. *RSC Advances*. 3, 8203–8207. Reproduced by permission of The Royal Society of Chemistry.)

substrate surface (during the IP process), a large amount of nanoparticles are removed together with the amine solution, leaving only a small amount of nanoparticles in the substrate pores. Therefore, nanoparticles, which are commonly hydrophilic, should be surface modified in order to make them more compatible with the organic phase. However, most research has preferred to disperse nanomaterials in amine aqueous solution and this could be mainly due to the difficulties in producing a homogenous nanomaterial-organic mixture. The high surface energy of inorganic nanomaterials (500–2000 mJ cm^{-2}) coupled with high inter-particle interactions are the main reasons causing them to aggregate easily in nature (Caseri, 2000).

It has been previously reported that instead of using hydrophilic nanoparticles, embedding PA layers with hydrophilic nanotubes such as aluminosilicate SWNTs, MWCNTs, and halloysite nanotubes (HNTs) could create a preferential pathway for water molecules, leading to an improvement in water flux (Amini et al., 2013; Baroña et al., 2013; Ghanbari et al., 2015). Nanotubes, however, have lengths between 10 and 50 μm (equivalent to 10,000 and 50,000 nm) (Wu et al., 2010; Kim et al., 2012; Amini et al., 2013; Shen et al., 2013). They cannot be accommodated in a thin PA layer of 100–500 nm (Lau et al., 2014) unless all of them are horizontally oriented, which is highly unlikely for the IP process because of the large negative entropy that cannot be compensated by the enthalpy term due to the poor interaction between the nanoparticles and PA. If this orientation indeed occurs, it is still contradictory to the claim of many researchers that water flows through vertically aligned channels. Nevertheless, randomly arranged nanotubes have to certain extent create a relatively less resistant pathway for water molecules to pass through, leading to flux improvement with minimum change in salt rejection as evidenced in several research works. Additional research needs to be carried out to verify the water and ion transport mechanism for these membranes, as improved flux has also been demonstrated for nonporous nanoparticles such as silver (Ag). Changes in free volume may be more relevant to the transport mechanism, and Xie et al. (2014) have shown free volume changes in poly(vinyl alcohol) (PVA) upon the addition of silica nanoparticles and related this to changes in permeability and selectivity. However, only few studies have examined such effects as gaining access to instruments capable of characterizing thin films in this manner is difficult.

On the other hand, the lack of chemical interaction between the nanomaterials and PA matrix is likely to cause the nanomaterials to easily leach out during the IP and/or filtration process, causing the resultant TFN membrane to be less effective and the efficiency of nanoparticles used during production to be low. Previous research works have always shown little problem with the incompatibility between the inorganic nanomaterials and organic PA layer. However, no thorough study has reportedly been carried out so far to evaluate nanoparticle leaching during a long period of filtration. Is the physical interaction sufficient to hold the nanomaterials in the PA layer especially under the high-pressure operation? What will be the outcome if the surface of nanomaterials is chemically bound to the PA structure? Will the chemical interaction be better for improving TFN performance stability over long-term operation? Efforts have been recently made to synthesize nanomaterials surface functionalized with $-NH_2$ groups in an attempt to improve PA-nanomaterials interaction (Amini et al., 2013; Wu et al., 2013; Peyravi et al., 2014; Emadzadeh et al., 2015), but in-depth analysis on the PA layer chemistry upon addition of functionalized nanomaterials has never been reported. Detailed study on the molecular interaction between PA and functionalized nanomaterials is worthy of special attention.

4.2.2 Innovative Approaches

Undoubtedly, the challenges in TFN membrane fabrication as mentioned above have motivated many dedicated scientific investigations in which some of the innovative approaches can be potentially employed to address these challenges.

4.2.2.1 Surface Modification of Nanomaterials

One of the simple yet effective approaches to improve the dispersion quality of nanomaterials in nonpolar organic solvent is through the surface modification of nanoparticles. The modified nanoparticles usually contain specific functional groups so as they can disperse homogenously/better in the nonpolar organic solution during the IP process to reduce the extent of particle agglomeration in the PA layer or to have chemical bonding with the PA network. Although surface deposition of nanoparticles could make the TFN membranes more active due to the high exposure of nanoparticles to membrane surface, the deposited nanoparticles might be lost during high-pressure filtration processes. Therefore, incorporating functionalized nanoparticles into the PA layer is much better to enhance the stability of nanoparticles in the TFN membrane surface due to the partial or full encapsulation of nanoparticles inside the PA thin layer.

As reported by Shen et al. (2013), surface modification of MWCNTs with acid solution (HNO_3 and H_2SO_4 mixture) followed by microemulsion polymerization of the methyl methacrylate monomer resulted in highly soluble nanomaterials in organic solvents. As shown in Figure 4.2, the synthesized polymethyl methacrylate (PMMA)–MWCNTs formed a very stable, uniform dark dispersion in toluene even after more than 1 month of standing. With this promising dispersion quality of nanomaterials, it is likely to minimize the precipitation of nanomaterials in organic solvent during IP and further to reduce the extent of particle agglomeration in the PA layer.

To assist proper dispersion of UZM-5 zeolite nanoparticles in organic (hexane) solvent, Namvar-Mahboub et al. (2014) functionalized the nanoparticles using an amino silane coupling agent—3-aminopropyldiethoxymethylsilane (APDEMS). It is claimed that the presence of the amino functional

FIGURE 4.2
Direct observation on the dispersion quality of PMMA–MWCNTs in toluene–water after 1 month. (Reprinted from *Journal of Membrane Science*, 442, Shen, J.N. et al., Preparation and characterization of thin-film nanocomposite membranes embedded with poly(methyl methacrylate) hydrophobic modified multiwalled carbon nanotubes by interfacial polymerization, 18–26, Copyright 2013, with permission from Elsevier.)

group (–NH_2) on the nanozeolite surface not only enables nanoparticles to disperse better in organic solvent but also possibly form a covalent bond with TMC molecules, improving the interaction of inorganic nanoparticles with the organic PA structure. The use of an amino silane coupling agent for modification of nanoparticles was also reported by Rajaeian et al. (2013) and Lai et al. (2016) in which they used N-[3-(trimethoxysilyl)propyl]ethylenediamine (AAPTS) to modify surface of titanium dioxide (TiO_2) nanoparticles and titanate nanotubes (TNTs), respectively. In the case of TiO_2 nanoparticles, the reduced oxygen bridge between the nanoparticles by AAPTS modification is said to be effective for minimization of the negative impact of particle agglomeration in the PA layer. However, the authors only demonstrated good dispersion of AAPTS-TiO_2 in the aqueous phase instead of the organic phase. Unlike Rajaeian et al. (2013), Lai et al. (2016) demonstrated the dispersion quality of TNTs with and without AAPTS modification in the organic solvent for up to 90 min. As can be seen from Figure 4.3, the AAPTS-modified TNTs were clearly dispersed better in the nonpolar solvent in comparison to the unmodified TNTs. Besides, the modified TNTs took a longer time to settle. Surface modification of MSN was conducted by Wu et al. (2013) using APS to prepare the TFN membrane for NF application.

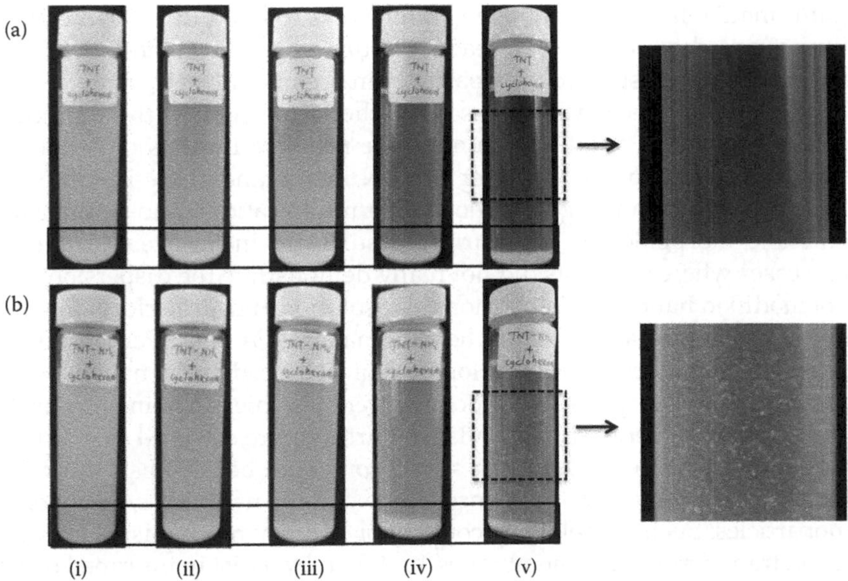

FIGURE 4.3
Dispersion quality of (a) TNTs and (b) AAPTS-modified TNTs in cyclohexane at different times, (i) 5 s, (ii) 2 min, (iii) 4 min, (iv) 10 min, and (v) 90 min. (Lai, G.S. et al. 2016. A practical approach to synthesize polyamide thin film nanocomposite (TFN) membranes with improved separation properties for water/wastewater treatment. *Journal of Materials Chemistry A*. 4, 4134–4144. Reproduced by permission of The Royal Society of Chemistry.)

Similar to Rajaeian et al. (2013), the authors did not report the dispersion of nanoparticles in the organic solvent. An explanation was given only on the possibility of chemical interaction between the functionalized silica nanoparticles and PA structure.

Peyravi et al. (2014) made an attempt to functionalize the surface of TiO_2 nanoparticles using two amine reagents, that is, MEOA and TETA. The aminated TiO_2 nanoparticles were claimed to provide a suitable surface for dispersion of nanoparticles in acyl chloride solution during the IP process. However, details were not provided for the dispersion quality of the functionalized nanoparticles in aqueous/organic solution. The effectiveness of this approach for TFN membrane fabrication was also not very convincing as the results from microscopy analysis revealed significant particle agglomeration on the PA top surface. On the other hand, a review article by Faure et al. (2013) had shown other potential chemical additives for improving the dispersion stability of TiO_2 nanoparticles in organic media, but these approaches have not been used for TFN membrane fabrication. For instance, polybutene–succinimide pentamine (OLOA 370) was proposed by Erdem et al. (2000) as a stabilizer to improve the dispersion quality of hydrophilic TiO_2 nanoparticles in cyclohexane. The amine end group of OLOA 370 is able to accept protons or donate an electron pair which results in a concentration-dependent negative zeta potential on the TiO_2 surface in low dielectric organic media. In other words, the amine end group on this polymeric stabilizer can readily interact with reactive hydroxyl group on the TiO_2 surface, leading to highly dispersed nanoparticles in organic solvent.

While researchers have proposed methods to modify the surface of nanoparticles in an effort to minimize the negative impacts of nanomaterial agglomeration in the PA layer of TFN membrane, there is still a lack of detailed understanding as to how such modifications lead to improved membrane morphologies and current results are inconsistent. There are many cases where scientists did not really demonstrate the dispersion quality of modified nanomaterials in nonpolar solution and/or perform in-depth analysis on the characteristics of the nanomaterial-PA layer. Perhaps, future research should focus on the development of advanced/novel nanomaterials that could readily disperse in nonpolar solvent to achieve a homogenous distribution of the materials in the PA layer. Furthermore, it would also be interesting if quantitative modeling of separation could be established for TFN membranes, particularly those incorporated with nanotubes/mesoporous nanoparticles. Such models may confirm or identify mechanisms for water and ion transport in TFN membranes, and thereby assist in informed design of better performing membranes.

4.2.2.2 Use of Metal Alkoxides

As solid hydrophilic nanoparticles are readily aggregated when mixed with nonpolar organic solvent, good dispersion of them in nonpolar solvent is difficult to achieve. To solve this problem, tetra isopropoxide (TTIP) which is readily dissolved in organic solvent was proposed to replace raw TiO_2 nanoparticles in preparing TFN membrane for RO application (Kong et al., 2011). Besides showing good dispersion in hexane, metal alkoxides could be hydrolyzed to produce smaller inorganic nanoparticles (and organic alcohol) either during or after the IP process. The results showed that in comparison to the typical TFC membrane, the water flux of TFN membrane was improved with only a slight decrease in salt rejection when a small amount of TTIP solution (~0.1 wt%) was added to the hexane solvent. However, when the performance of the TTIP-TFN membrane was further compared with the TFN membrane prepared from other metal alkoxides—phenyl triethoxysilane (PhTES), it was found that the PhTES-TFN membrane was much better as the latter membrane could overcome the trade-off effect between water permeability and solute rejection. The reasons for these variations in performance, however, were not identified by the authors. The potential of using metal alkoxides in TFN membrane fabrication is still not very clear as there is no other relevant article in the open literature.

4.2.2.3 Modified/Novel IP Techniques

In terms of the IP process, Lau et al. (2012) revealed the continuous concerted efforts made by researchers in preparing TFC membranes with improved interfacial properties via a modified IP procedure. Some of these modifications are the introduction of a secondary amine solution to react again with unreacted acyl chloride groups by placing the substrate membrane in an aqueous solution for a second time (Zou et al., 2010) or employment of an intermediate organic solvent between the aqueous amine solution and the organic acid chloride solution so as to reduce possible pinhole formation on the PA layer (Veríssimo et al., 2005). A further review is conducted in this section to explore how a modified IP technique could be possibly adopted for a composite membrane, in particular the PA TFN membrane fabrication process.

Kong et al. (2010) introduced an intermediate "pre-seeding" hexane solution containing low concentration of TMC (0.02–0.05 wt% TMC, 0.05–0.6 wt% zeolite, and 5–15 wt% ethanol) between 2 wt% MPD aqueous solution and 0.1 wt% TMC organic solution to assist in the dispersion of zeolite nanoparticles on the top surface of the PSF substrate, as illustrated in Figure 4.4a. The field emission scanning electron microscopy (FESEM) images (see Figure 4.4b and c) clearly showed well-dispersed zeolite nanoparticles throughout the top PSF substrate by using a "pre-seeding" solution containing 0.4 wt% zeolite. With respect to filtration performance, the TFN membrane fabricated by a "pre-seeding" method with 0.2 wt% zeolite loading showed greater water permeability of 4.2×10^{-12} m/Pa.s in comparison to 2.9 and 1.9×10^{-12} m/Pa.s recorded in the TFN membrane prepared by adding zeolite into TMC-hexane and MPD aqueous solution, respectively. Even though the water permeability of

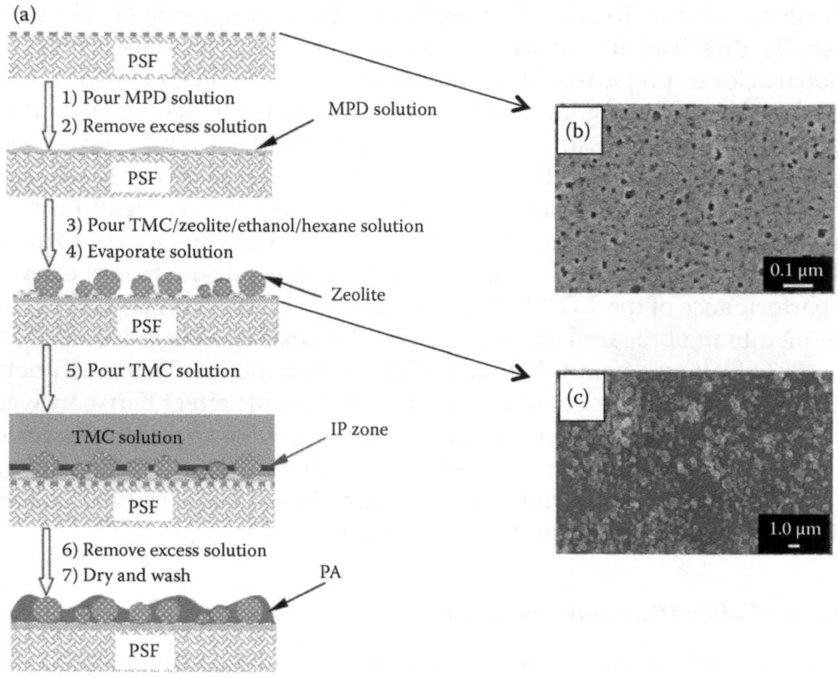

FIGURE 4.4
(a) Schematic representation of the TFN membrane fabrication via "pre-seeding" method and FESEM images of PSF support surface before (b) and after (c) "pre-seeding" process. (Kong, C., Shintani, T., Tsuru, T. 2010. "Pre-seeding"-assisted synthesis of a high performance polyamide-zeolite nanocomposite membrane for water purification. *New Journal of Chemistry*. 34, 2101–2104. Reproduced by permission of The Royal Society of Chemistry.)

TFN membrane prepared by the modified IP was remarkably improved, high NaCl rejection (97.4%) was not compromised. The promising results were attributed to the well-defined zeolite pores that serve as a short cut for water permeation. Rejection experiments conducted using neutral organic solutes of different molecular sizes showed that the TFN membrane prepared exhibited very similar "pore" size (~0.79 nm) in comparison to the pore size of incorporated zeolite Y (~0.74 nm), suggesting the pore channels of nanoparticles were still accessible for transport following the IP process.

A dynamic IP process illustrated in Figure 4.5 was used by An et al. (2012) to form a PA layer on a microporous substrate. This dynamic IP was employed to fabricate TFC membrane for pervaporation process, and the results were quite interesting with respect to the surface pattern of the PA layer. The powerful centrifugal force created during the spinning process tends to align molecular chains in the growing interfacial PA film in the horizontal direction, leading to the formation of a thinner and smaller free volume PA structure. The uniquely formed PA structure was verified by PAS in which the lower value of the *S* parameter was obtained with an increasing

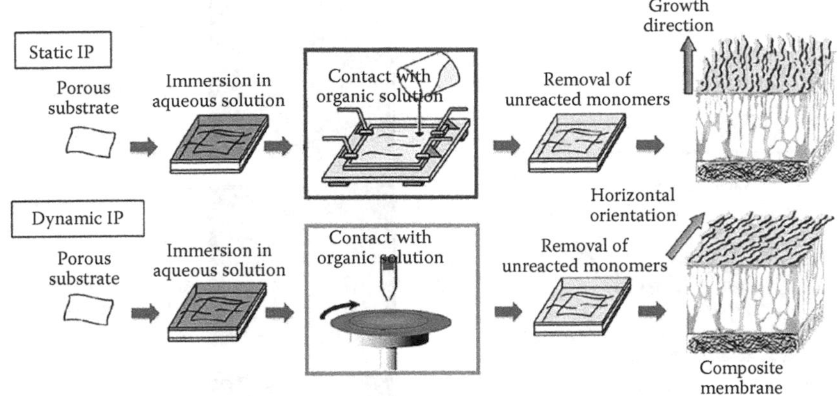

FIGURE 4.5
Establishing PA layer on top of microporous substrate through static and dynamic IP. (Reprinted with permission from An, Q. et al. 2012. Comparison between free volume characteristics of composite membranes fabricated through static and dynamic interfacial polymerization processes. *Macromolecules*. 45, 3428–3435. Copyright 2012 American Chemical Society.)

spin coating rate. In terms of performance, the PA film formed from applying the dynamic IP process showed simultaneous improvement in flux and selectivity, overcoming the trade-off phenomenon that is commonly known in the conventional IP process (static). Although this novel IP process is only attempted for TFC membrane making, it could be possibly used for TFN membrane preparation as well (Fu et al., 2014). It is because the centrifugal force induced by a spin coater is likely to spread the organic solution (which contains nanoparticles) from the axis of rotation toward the outer substrate edge, leading to good dispersion of the nanoparticles over the entire substrate surface. In addition, the advantage of spin coating in forming an ultrathin PA selective layer (around 200 nm) might potentially reduce water transport resistance during the filtration process and enhance membrane water productivity. This fabrication approach is one of the subjects deserving greater research attention.

4.2.2.4 Alignment of Nanotubes/Fillers

Although there is not any research work conducted so far on alignment of nanotubes in composite membranes prepared via the IP process, one can still find several innovative approaches that have been recently employed for asymmetric mixed matrix membrane fabrication to control nanofiller orientation either vertically or horizontally in the membrane structure. Alignment of nanofillers, for example, CNTs, in a membrane matrix could reduce interfacial voids between the CNTs and the polymer, minimizing discontinuous and tortuous paths for water molecules to be transported.

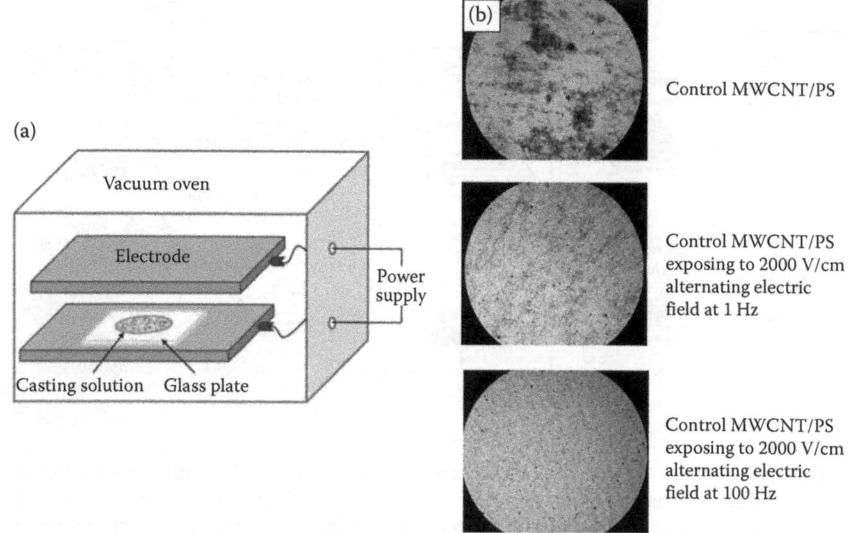

FIGURE 4.6
(a) Schematic of an apparatus for electrical alignment of MWCNTs in polymer matrix. (b) Micrographs of 0.3 wt% MWCNTs dispersed in PS membrane with and without applying electric field. (Reprinted from *Journal of Membrane Science*, 462, Wu, B. et al., Electro-casting aligned MWCNTs/polystyrene composite membranes for enhanced gas separation performance, 62–68, Copyright 2014, with permission from Elsevier.)

Wu et al. (2014) reported a new method to align MWCNTs in polystyrene (PS) membrane by means of an alternating electric field to achieve an even dispersion of MWCNTs. Figure 4.6 shows a schematic diagram of the electrical field alignment apparatus used for MWCNT/PS membrane preparation and the micrographs of the membranes prepared from 3 wt% MWCNTs at different alternating electric field frequencies. As can be seen, the extent of MWCNTs' aggregation and maldistribution is significantly reduced in the electro-cast membranes. Increasing the frequency of the electric field from 1 to 100 Hz was also found to further improve dispersion of CNTs in the membrane matrix. These results can be explained by the electric field which aligns the conductive MWCNTs in the direction of the field and causes them to exclude each other perpendicularly via dipole–dipole interactions. Because of this, the electro-cast membranes showed much improved oxygen permeabilities over the control membranes.

Kim et al. (2014) developed a novel *in situ* bulk polymerization method to prepare vertically aligned carbon nanotube (VACNT)/polymer composite membrane for both gas and water separation processes. In order to prevent CNT condensation that could disturb CNT orientation during liquid-phase processing, the VACNT array was infiltrated with a styrene monomer with a certain amount of polystyrene–polybutadiene block copolymer that acts as a plasticizer. SEM images identified that well-aligned CNTs were embedded in a high-density polymer matrix free of any macroscopic voids or structural defects.

In addition to these two possible alignment approaches, a review article written by Goh et al. (2014) have summarized various contemporary approaches that could be employed to align CNTs either horizontally or vertically in the polymer matrix. However, more research is needed to determine if these approaches could be used for the thin PA layer synthesis process or are only limited to asymmetric membrane structure.

References

Amini, M., Jahanshahi, M., Rahimpour, A. 2013. Synthesis of novel thin film nanocomposite (TFN) forward osmosis membranes using functionalized multiwalled carbon nanotubes. *Journal of Membrane Science*. 435, 233–241.

An, Q., Hung, W., Lo, S., Li, Y., De Guzman, M. 2012. Comparison between free volume characteristics of composite membranes fabricated through static and dynamic interfacial polymerization processes. *Macromolecules*. 45, 3428–3435.

Baroña, G.N.B., Lim, J., Choi, M., Jung, B. 2013. Interfacial polymerization of polyamide–aluminosilicate SWNT nanocomposite membranes for reverse osmosis. *Desalination*. 325, 138–147.

Caseri, W. 2000. Nanocomposites of polymers and metals or semiconductors: Historical background and optical properties. *Macromolecular Rapid Communications*. 21, 705–722.

Emadzadeh, D., Lau, W.J., Rahbari-Sisakht, M., Daneshfar, A., Ghanbari, M., Mayahi, A., Matsuura, T., Ismail, A.F. 2015. A novel thin film nanocomposite reverse osmosis membrane with superior anti-organic fouling affinity for water desalination. *Desalination*. 368, 106–113.

Erdem, B., Sudol, E.D., Dimonie, V.L., El-Aasser, M.S. 2000. Encapsulation of inorganic particles via miniemulsion polymerization. I. Dispersion of titanium dioxide particles in organic media using OLOA 370 as stabilizer. *Journal of Polymer Science Part A: Polymer Chemistry*. 38, 4419–4430.

Faure, B., Salazar-Alvarez, G., Ahniyaz, A., Villaluenga, I., Berriozabal, G., De Miguel, Y.R., Bergström, L. 2013. Dispersion and surface functionalization of oxide nanoparticles for transparent photocatalytic and UV-protecting coatings and sunscreens. *Science and Technology of Advanced Materials*. 14, 1–23.

Fu, Q., Wong, E.H.H., Kim, J., Scofield, J.M.P., Gurr, P.A., Kentish, S.E., Qiao, G.G. 2014. The effect of soft nanoparticles morphologies on thin film composite membrane performance. *Journal of Material Chemistry A*. 2, 17751–17756.

Ghanbari, M., Emadzadeh, D., Lau, W.J., Lai, S.O., Matsuura, T., Ismail, A.F. 2015. Synthesis and characterization of novel thin film nanocomposite (TFN) membranes embedded with halloysite nanotubes (HNTs) for water desalination, *Desalination*. 358, 33–41.

Goh, P.S., Ismail, A.F., Ng, B.C. 2014. Directional alignment of carbon nanotubes in polymer matrices: Contemporary approaches and future advances. *Composites Part A: Applied Science and Manufacturing*. 56, 103–126.

Huang, H., Qu, X., Dong, H., Zhang, L., Chen, H. 2013. Role of NaA zeolites in the interfacial polymerization process towards a polyamide nanocomposite reverse osmosis membrane. *RSC Advances*. 3, 8203–8207.

Kim, E., Hwang, G., El-Din, M.G., Liu, Y. 2012. Development of nanosilver and multi-walled carbon nanotubes thin-film nanocomposite membrane for enhanced water treatment. *Journal of Membrane Science*. 394–395, 37–48.

Kim, S., Fornasiero, F., Park, H.G., In, J.B., Meshot, E., Giraldo, G., Stadermann, M. et al. 2014. Fabrication of flexible, aligned carbon nanotube/polymer composite membranes by in-situ polymerization. *Journal of Membrane Science*. 460, 91–98.

Kong, C., Koushima, A., Kamada, T., Shintani, T., Kanezashi, M., Yoshioka, T., Tsuru, T. 2011. Enhanced performance of inorganic-polyamide nanocomposite membranes prepared by metal-alkoxide-assisted interfacial polymerization. *Journal of Membrane Science*. 366, 382–388.

Kong, C., Shintani, T., Tsuru, T. 2010. "Pre-seeding"-assisted synthesis of a high performance polyamide-zeolite nanocomposite membrane for water purification. *New Journal of Chemistry*. 34, 2101–2104.

Kumano, A., Ogura, H., Hayashi, T. 1998. Composite hollow fiber membrane and process for its production. US Patent No. 5783079.

Lai, G.S., Lau, W.J., Gray, S.R., Matsuura, T., Jamshidi Gohari, R., Subramanian, M.N., Lai, S.O. et al. 2016. A practical approach to synthesize polyamide thin film nanocomposite (TFN) membranes with improved separation properties for water/wastewater treatment. *Journal of Materials Chemistry A*. 4, 4134–4144.

Lau, W.J., Ismail, A.F., Goh, P.S., Hilal, N., Ooi, B.S. 2014. Characterization methods of thin film composite nanofiltration membranes. *Separation and Purification Reviews*. 44, 135–156.

Lau, W.J., Ismail, A.F., Misdan, N., Kassim, M.A. 2012. A recent progress in thin film composite membrane: A review. *Desalination*. 287, 190–199.

Li, X., Chung, T.-S. 2014. Thin-film composite P84 co-polyimide hollow fiber membranes for osmotic power generation. *Applied Energy*. 114, 600–610.

Namvar-Mahboub, M., Pakizeh, M., Davari, S. 2014. Preparation and characterization of UZM-5/polyamide thin film nanocomposite membrane for dewaxing solvent recovery. *Journal of Membrane Science*. 459, 22–32.

Peyravi, M., Jahanshahi, M., Rahimpour, A., Javadi, A., Hajavi, S. 2014. Novel thin film nanocomposite membranes incorporated with functionalized TiO_2 nanoparticles for organic solvent nanofiltration. *Chemical Engineering Journal*. 241, 155–166.

Rajaeian, B., Rahimpour, A., Tade, M.O., Liu, S. 2013. Fabrication and characterization of polyamide thin film nanocomposite (TFN) nanofiltration membrane impregnated with TiO_2 nanoparticles. *Desalination*. 313, 176–188.

Shao, L., Cheng, X.Q., Liu, Y., Quan, S., Ma, J., Zhao, S.Z., Wang, K.Y. 2013. Newly developed nanofiltration (NF) composite membranes by interfacial polymerization for Safranin O and Aniline blue removal. *Journal of Membrane Science*. 430, 96–105.

Shen, J.N., Yu, C.C., Ruan, H.M., Gao, C.J., Van der Bruggen, B. 2013. Preparation and characterization of thin-film nanocomposite membranes embedded with poly(methyl methacrylate) hydrophobic modified multiwalled carbon nanotubes by interfacial polymerization. *Journal of Membrane Science*. 442, 18–26.

Sukitpaneenit, P., Chung, T.-S. 2014. Fabrication and use of hollow fiber thin film composite membranes for ethanol dehydration, *Journal of Membrane Science.* 450, 124–137.

Tadros, S.E., Trehu, Y.M. 1990. Coating process for composite reverse osmosis membranes. US Patent No. 4980061.

Veríssimo, S., Peinemann, K.-V., Bordado, J. 2005. New composite hollow fiber membrane for nanofiltration. *Desalination.* 184, 1–11.

Wu, B., Li, X., An, D., Zhao, S., Wang, Y. 2014. Electro-casting aligned MWCNTs/polystyrene composite membranes for enhanced gas separation performance. *Journal of Membrane Science.* 462, 62–68.

Wu, H., Tang, B., Wu, P. 2010. MWNTs/polyester thin film nanocomposite membrane: An approach to overcome the trade-off effect between permeability and selectivity. *The Journal of Physical Chemistry C.* 114, 16395–16400.

Wu, H., Tang, B., Wu, P. 2013. Optimizing polyamide thin film composite membrane covalently bonded with modified mesoporous silica nanoparticles. *Journal of Membrane Science.* 428, 341–348.

Xie, Z., Hoang, M., Ng, D., Doherty, C., Hill, A., Gray, S. 2014. Effect of heat treatment on pervaporation separation of aqueous salt solution using hybrid PVA/MA/TEOS membrane. *Separation and Purification Technology.* 127, 10–17.

Yang, F., Zhang, S., Yang, D., Jian, X. 2007. Preparation and characterization of polypiperazine amide/PPESK hollow fiber composite nanofiltration membrane. *Journal of Membrane Science.* 301, 85–92.

Yin, J., Kim, E.-S., Yang, J., Deng, B. 2012. Fabrication of a novel thin-film nanocomposite (TFN) membrane containing MCM-41 silica nanoparticles (NPs) for water purification. *Journal of Membrane Science.* 423–424, 238–246.

Zou, H., Jin, Y., Yang, J., Dai, H., Yu, X., Xu, J. 2010. Synthesis and characterization of thin film composite reverse osmosis membranes via novel interfacial polymerization approach. *Separation and Purification Technology.* 72, 256–262.

5

Characterization of Nanofiltration Membrane

5.1 Overview of NF Characterization

Since NF membranes, in particular interfacially polymerized membranes are different in several aspects such as material, morphology, transport/separation mechanism, and applications, a large number of different techniques are required for their characterizations. Characterization methods of NF membranes can be generally divided into three major groups, that is, methods used for chemistry evaluation, methods used for physical properties analysis, and filtration process for assessing membrane separation performance. Depending on the practical use of NF membranes, membrane stability tests against chlorination, organic solvent, thermal, and fouling can also be performed to examine their sustainability under specific environments.

Although brief descriptions of the analytical methods used to evaluate membrane properties and performances can be generally found in a number of reference books, in the past and present, the main highlight, and hence the most significant contribution of this chapter is to place special focus and emphasis on the characterization methods used for polymeric NF membranes. An attempt has been made to cover most of the NF characterization methods that also reflect the particular interests of the researchers. In order to ensure and facilitate better understanding of the discussed characterization methods in an easier way, the content in this chapter is organized into three main subsections. The most comprehensive one is concerned with the analytical instruments used for the chemical and physical properties characterization of NF membranes and thereafter the filtration process used to characterize NF permeability and selectivity. For both chemical and physical characterizations, a brief description of the measurement procedures is provided before presenting the expected outcomes from the respective analyses. Some practical advice is provided in the respective subsections with the aim of minimizing characterization error and improving the accuracy of data. The assessment methods applied on NF membrane stabilities are addressed

TABLE 5.1

Assessments on Membrane Properties and Performances Based on Different Analytical Instruments/Methods

Properties Assessment	Instrument/Method	Properties Assessment	Instrument/Method
Chemical properties	ATR-FTIR spectroscopy Zeta potential analysis XPS X-ray diffractometry (XRD) Nuclear magnetic resonance (NMR) spectroscopy	Physical properties	SEM/FESEM TEM Atomic force microscopy (AFM) Contact angle analysis PAS
Separation performance	Permeability Selectivity	Stability test	Chlorination Solvent Thermal Filtration

in the final subsection. There are four major stability tests for NF membranes under different process conditions, depending on their industrial applications. The stability tests can be classified into chlorination exposure, thermal resistance, solvent resistance, and long-term filtration process.

Table 5.1 lists the analytical instruments/methods that are commonly used in characterizing NF membranes with respect to their chemical and physical properties as well as separation performances and stability. In general, before conducting NF filtration experiments, various techniques can be employed for the characterization of the NF membranes in order to obtain a good knowledge of the different membrane parameters that are important for producing a membrane with the right combination of water flux and solute rejection.

5.2 Instruments/Methods for Chemical Properties Assessment

5.2.1 Attenuated Total Reflectance-FTIR Spectroscopy

In order to analyze functional groups and bonds of TFC NF membranes prepared from various monomer combination and substrates, attenuated total reflectance (ATR) is a popular sampling technique used in conjunction with infrared (IR) spectroscopy. The combined techniques enable membrane samples to be examined directly without further preparation. The IR spectra reveal not only the band ascribable to the PA layer but also for those due to the substrate as the IR beam penetration depth exceeds the typical thickness of the PA layer. The typical depth of the reflected IR beam penetration in the ATR technique is around 1 μm (Wang et al., 2010) while the thickness of the PA layer is several hundreds of nm (Jadav and Singh 2009). In order to show only the spectra of the PA layer, the substrate of the TFC membrane can be digitally subtracted. Other than this, there is a method to separate the dense selective layer from the microporous substrate by subjecting the TFC membrane to a chloroform solution prior to FTIR analysis (Barona et al., 2013). With this approach, the substrate that is made of either PSF or PES will be dissolved in the chloroform solution and a PA layer can be individually obtained.

Prior to FTIR analysis, it is recommended to dry membrane samples in a desiccator for a minimum of 24 h and to purge the instrument with nitrogen to prevent interference of atmospheric moisture with the spectra. Dry and clean samples are prepared and pressed directly on the crystal made of either germanium (Ge) or zinc selenide (ZnSe) during measurement. An internal reflection element with incidence angle of 45° is often employed. For each IR study conducted on an NF membrane, the spectrum can be evaluated in the range from 1100 to 3600 cm^{-1} with at least 32 scans (preferably 64 scans) at a resolution of 4 cm^{-1}.

Table 5.2 summarizes the major peak characteristics detected from a PA layer. The presence of amide I, amide II, and aromatic ring at 1620–1662, 1537–1575, and 1600–1610 cm^{-1}, respectively, is the evidence of the formation of functional —NHCO— bond (acrylamide group). Peak assignment due to amide II arises from the couplings of in-plane N—H bending and C—N stretching vibration of the C—N—H group. Instead of the peaks shown by PA layer, the absence of acyl chloride (monomer) peak at 1760–1770 cm^{-1} can also indicate the occurrence of IP. In the case where secondary diamine—piperazine (PIP) is used as the sole amine monomer during the IP process, one cannot observe the peak at 1540–1554 cm^{-1}. This is due to the absence of N—H bond in the amide formed with acid chloride (—RCON—) (Saha and Joshi, 2009).

In recent years, research on developing a new generation of hybrid top selective layer (by incorporating inorganic nanomaterials into the PA layer) has become the popular topic of study among membrane scientists with the

TABLE 5.2

Major Peak Characteristics of PA Layer

Peak Assignment	Wavenumber (cm^{-1})	Band Signal
Amide I, C=O stretching	1620–1662	Strong
Amide II, N—H bending	1540–1554	Strong
C—N stretching	1537–1575	Strong
Aromatic ring breathing, C_6H_6	1600–1610	Strong
Carboxylic acid, O—H stretching	3300–3410	Weak
O—H bending	~1450	Weak

aim of overcoming the trade-off effect between water permeability and solute selectivity. To evaluate the presence of inorganic nanofillers in the PA layer, one can also perform the analysis using FTIR spectroscopy. For instance, the bands at around 654 and around 1040 cm^{-1} suggest the stretching vibration and anti-stretching vibration of Si–O band upon the addition of silicon dioxide (SiO_2) nanoparticles into the PA layer (Jadav and Singh, 2009; Hu et al., 2012; Yin et al., 2012). A wide peak at about 900–1200 cm^{-1} indicates the presence of zeolite NaX in the PA matrix (Fathizadeh et al., 2011). Wu et al. (2013a), on the other hand, characterized the formation of the amide bond between mMSN with amino groups and TMC using the FTIR technique. However, owing to the severe overlaps of the vibration band in the IR spectrum of the resultant composite membrane, the chemical reaction between mMSN and TMC was able to only be analyzed based on the mMSN–TMC compound. This compound was obtained by directly adding mMSN into the TMC/cyclohexane solution followed by washing and drying.

Apart from the peaks displayed by the PA layer, Table 5.3 shows the functional groups of some commonly used microporous substrates together with their respective wavenumbers. Among all the polymers used, PSF and PES are the most commonly used materials both in the past and present for

TABLE 5.3

Peak Characteristics of Typical Microporous Substrate Blended with and without Additives

Peak Assignment	Wavenumber (cm^{-1})
Pure PES substrate	
C—O—C stretching (asymmetric)	~1250
CH_3—C—CH_3 stretching	~1500
Benzene ring	~1578
Pure PSF substrate	
O=S=O stretching (symmetric)	~1150
O=S=O stretching (asymmetric)	1290 and 1325
C—O—C stretching (asymmetric)	~1250
CH_3—C—CH_3 stretching	~1500
Aromatic C—C stretching	~1600
Pure PEI substrate	
C=O stretching (asymmetric)	~1780
C=O stretching (symmetric)	~1720
C—N stretching	~1350
Presence of additive in substrate	
(a) PEG	
Ether (R—O—R)	2900–3000
Hydroxyl (R—OH)	2900–3000
(b) PVP	
Amide group	~1650

substrate preparation for in-house manufactured and commercial TFC NF membranes. The organic structures of PSF and PES are very similar, except PES possesses a higher concentration of sulfone moieties in the polymer repeat unit. Other polymers such as PEI and PAN are also considered for substrate making, but only reported in laboratory studies.

Instead of using substrate made of a single polymer, the use of additives to increase substrate porosity and/or hydrophilicity is also generally practiced. With the addition of a small amount of hydrophilic polyvinylpyrrolidone (PVP) (preferable 1 wt%) into the substrate matrix, a new significant peak at around 1650 cm^{-1} assigned to the primary amide stretch can be observed. In contrast to the substrate blended with PEG, it is hard to see any difference compared to the pure substrate. The main reason for this result is attributed to the overlapping between the bands of PEG and substrate (Susanto and Ulbricht, 2009) and/or the possible PEG leaching out from the substrate during the phase inversion process. The detection of PVP in the substrate matrix is also not definite as the PVP is likely to leach out from the substrate matrix during the phase inversion process due to its hydrophilic nature, causing very little or none of it remaining in the substrate.

5.2.2 Zeta Potential

Zeta potential is a parameter used to characterize the surface charge property of NF membranes at different pH environments. The analysis is particularly important to help understand the acid–base properties of NF membranes, and to predict the separation efficiency, as well as to analyze the fouling propensity of NF at different water pHs. On the basis of the Helmholtz–Smoluchowski equation with the Fairbrother and Mastin approach, zeta potential, ζ, can be determined from measureable streaming potential using Equation 5.1:

$$\zeta = \frac{\Delta E}{\Delta P} \frac{\mu \kappa}{\varepsilon \varepsilon_0} \tag{5.1}$$

where ΔE is the streaming potential, ΔP is the applied pressure, μ is the solution viscosity, κ is the solution conductivity, ϵ and ϵ_0 are the permittivity of the test solution and free space, respectively. Several assumptions are inherent in this equation. They are (1) flow is laminar, (2) surface conductivity has no effect and has homogeneous properties, (3) width of the flow channel is much larger than the thickness of the electric double layer, and (4) no axial concentration gradient occurs in the flow channel.

A membrane sample is required to be soaked in pure water or potassium chloride (KCl) electrolyte solution for a day to remove unreacted monomer off the membrane surface prior to sample analysis. Measurement can then be performed using either a commercial electrokinetic analyzer or self-assembled experimental setup as described in the work of Liu et al. (2009). Since only the surface (PA dense selective layer) of the NF membrane is analyzed, the adjustable gap cell or clamping cell is the appropriate device to be used. During sample measurement, a streaming potential is induced when ions within an electrical double layer are forced to move along with a flow tangential to the membrane surface, hence resulting in a potential difference to be generated. It must be pointed out that for the TFC NF membrane; it is the streaming potential of the membrane surface, not the membrane pores, that is being measured. The typical pH range applied for determining surface zeta potential of an NF membrane used to fall within pH 2–12 (more preferably, it is between pH 3 and 9). The pH of the background electrolyte solution (5 mM KCl, 25°C) can be adjusted through the addition of either 0.1 M HCl (or HNO_3) or 0.1 M NaOH (or KOH) solution. Owing to the possible irreversible change of membrane property following strong acid/alkali exposure, it is highly recommended to conduct this experiment using two identical fresh membranes, that is, one for acid titration (from pH 6 down to pH 2) followed by another identical membrane for alkali titration (from pH 6 up to pH 12).

Figure 5.1 presents the typical surface zeta potential of NF membranes measured at different pH environments (Liu et al., 2012a). Clearly, at lower pHs (<pH 3.5), the NF membranes tended to have more positive charge owing to the protonation of the amine functional groups. In contrast, the

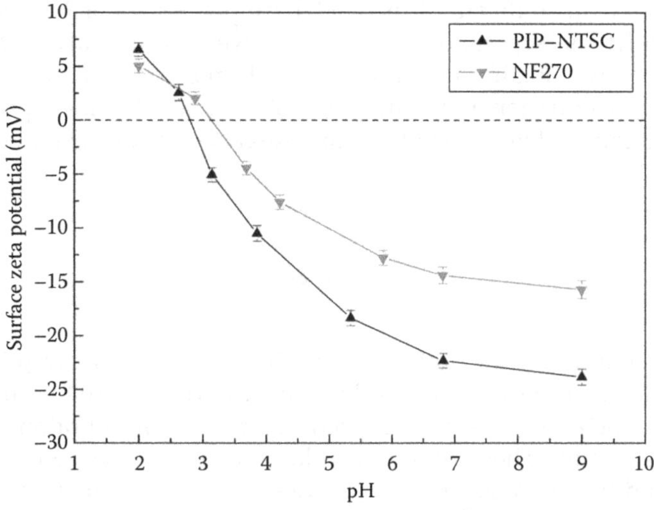

FIGURE 5.1
Surface zeta potential as a function of pH for composite membrane made of PIP and NTSC and commercial NF270 membrane manufactured by DOW FILMTEC. (Reprinted from *Journal of Membrane Science*, 415–416, Liu, M. et al., Acid stable thin-film composite membrane for nanofiltration prepared from naphthalene-1,3,6-trisulfonylchloride (NTSC) and piperazine (PIP), 122–131, Copyright 2012, with permission from Elsevier.)

negative charge of NF membranes at higher pHs can be attributed to the deprotonation of amine functional groups coupled with either dissociation of the carboxylic acid group (in NF270 membrane) or sulfonic acid group (in PIP–naphthalene-1,3,6-trisulfonylchloride (NTSC) membrane) on the membrane surface. In brief, the PIP–NTSC membrane made of a novel monomer is more negatively charged than that of the commercial NF (PIP–TMC) at pH 7. Besides showing the positive and negative charge value of a membrane, zeta potential profile can also reveal the IEP of the NF membrane at which the membrane surface carries no net electrical charge (i.e., neutral).

Depending on the functional groups of NF surface, a highly positively charged NF membrane could also be prepared in which this membrane displays a positive zeta potential over a wide range of pH values (pH 2–11) (Zhang et al., 2011). The phenomenon is mainly due to the presence of pendant tertiary amine groups in the polymer used to fabricate the membrane. The existence of ammonium form ($-NH_3^+$) in the PA selective layer upon the interaction of a novel monomer (PEG600–NH_2) with TMC was also reported to cause the membrane to be positively charged for pH ranging from 3 to 9 (Cheng et al., 2015). A summary of the surface zeta potential of some NF membranes made of different monomers at two different pH environments is presented in Table 5.4.

It should be noted here that besides surface zeta potential measurement, the conventional titration method can also be employed to evaluate the ion exchange capacity of the NF membrane. Any change in the membrane ion exchange capacity can be related to the amount of charged groups that exist on a membrane (Schaep and Vandecasteele, 2001; Schafer et al., 2005). Schaep and Vandecasteele (2001) for instance determined the charge of four commercially available NF membranes using the titration method in which a cesium chloride (CsCl) and sodium fluoride (NaF) solution was used to measure the negatively and positively charged groups of NF, respectively. However, it was reported that the results obtained from the titration method were hard to compare quantitatively with zeta potential measurement as each method has a different approach.

5.2.3 X-Ray Photoelectron Spectroscopy

XPS is a quantitative spectroscopic method that is helpful in providing information on the elemental composition of the PA layer on the composite membrane. One is able to use this information to obtain insights into the likely CL structure of the PA layer. As ATR-FTIR is able to provide significant qualitative details about the types of functional groups present in the PA layer, it is generally considered as a highly complementary method to XPS.

In principle, XPS is conducted according to the photoelectric effect in which electrons are emitted from a surface after it adsorbs energy from light. Three main elements—carbon (C), nitrogen (N), and oxygen (O) are often selected to be quantified during XPS analysis as the presence of these

TABLE 5.4
Summary of the Surface Zeta Potential of Some NF Membranes at Different pH Environments

Type of NF Membrane[a]	IEP (pH)	ζ (mV) at pH 3	ζ (mV) at pH 9
MPF-34 (Koch Membrane Systems) (Dalwani et al., 2011)	4.5	~13	~ –34
Desal-5DK (GE Osmonic) (Dalwani et al., 2011)	3.9	~18	~ –50
NF 270 (DOW FILMTEC) (Dalwani et al., 2011)	3.2	~5	~ –75
BW30 (DOW FILMTEC) (Kwon et al., 2012)	3.6	~2	~ –10
NF90 (DOW FILMTEC) (Kwon et al., 2012)	4.2	~14	~ –24
PIP–TMC–MWCNT NF membrane (Shen et al., 2013)	2.6	~ –1.2	~ –7
MPD–TMC NF membrane (Wang et al., 2010)	6.0	~28	~ –11
PIP–TMC–GO NF membrane (Lai et al., 2016)	5.4	~25	~ –32
PVAm–TMC NF membrane (Liu et al., 2012b)	6.5	~19	~ –12
AEPPS–PIP–TMC NF membrane (An et al., 2013)	4.1	~1.3	~5.6
PES–TA NF membrane (Zhang et al., 2011)[b]	10.7	~32	~6
PIP–*mm*-BTEC NF membrane (Wang et al., 2011)[b]	–	~28	~4
PEG600–NH_2–TMC NF membrane (Chen et al., 2015)[b]	~8.9	~19	0

[a] AEPPS—N-aminoethyl piperazine propane sulfonate, MPD—m-phenylenediamine, *mm*-BTEC—3,3′,5,5′-biphenyl tetraacyl chloride, MWCNT—multi-walled carbon nanotube, GO—graphene oxide, PES–TA—poly(arylene ether sulfone) with pendant tertiary groups, PIP—piperazine, PVAm—polyvinylamine, and TMC—trimesoyl chloride.

[b] These NF membranes are positively charged over a wide pH range.

elemental compositions at ~285, ~400, and ~530 eV, respectively are strongly linked to the degree of CL in the PA thin layer. The spectra of other elements might also be considered when a specific component is incorporated into the aromatic PA layer during the IP process. For instance, the presence of aluminosilicate SWNTs in the thin film layer could be analyzed with respect to aluminum (Al) and silicon (Si) at XPS spectrum of ~75 and ~102.5 eV, respectively (Barona et al., 2012). Titanium dioxide (TiO_2) nanoparticles meanwhile could be detected at around 453.8 and 460 eV (Lee et al., 2008). Nevertheless, XPS measurement is a surface measurement that probably penetrates 2–10 nm below the PA surface. Because of this limitation, the results might not be accurate in the cases of PA-nanocomposite layers where nanofillers are usually embedded beneath the top PA layer.

For XPS analysis, Al/Kα (h_v = 1486.6 eV) or Mg/Kα (h_v = 1253.6 eV) is often the photon energies of choice. Other X-ray sources such as Ti/Kα (h_v = 2040 eV) can also be chosen. During measurement, the spectra are taken with the electron emission angle at 54° to give a sampling depth of 10 nm. A concentric hemispherical energy electron analyzer which operates at a constant pass energy of 29.35 eV is typically employed for a narrow scan. To facilitate transmission of the photoelectrons to the analyzer, the sample analysis is conducted in a vacuum chamber, under ultrahigh vacuum conditions, typically 9–10 mbar.

Table 5.5 compares the relative surface atomic concentrations and atomic ratio of the NF membranes obtained from manual calculation and XPS analysis (Wang et al., 2011). As can be seen, the reaction of PIP with 3,3′,5,5′-biphenyl tetraacyl chloride (*mm*-BTEC) could produce five different possible chemical structures owing to the multifunctional monomers used. By comparing the results obtained from XPS with manual calculation, one is able to confirm which chemical structure is most likely to exist in the PA layer. Since the atomic ratio of O/N as measured by XPS is very close to the ratio of O/N determined in unit C proposed, it can be said that the PA layer possesses more terminal amine functional groups. The presence of terminal amine functional groups is also believed to cause the NF membrane to have positive surface charge for a wide range of pH (pH 3–10) as confirmed by the zeta potential measurement. As shown in Table 5.5, it is highly possible that both cross-linked and linear parts coexist in the real PA backbone. Although O/N atomic ratio is the most common criterion for investigating the extent of PA thin film CL, the use of this ratio may cause error in the reported values if inorganic nanomaterials such as zeolite and titanium nanotubes are embedded in the PA layer (for the case of TFN membrane) since the addition of these nanomaterials could introduce extra oxygen to the system and affect the O/N ratio. Because of this, the C/N ratio is considered for evaluating the film CL. To determine the degree of PA CL based on C/N ratio, Equation 5.2 can be employed (Barona et al., 2013):

$$\mathrm{CL} = \left[1 - \left(\frac{(\mathrm{C/N})_{\text{observed}} - (\mathrm{C/N})_{\text{fully cross-linked}}}{(\mathrm{C/N})_{\text{fully linear}} - (\mathrm{C/N})_{\text{fully cross-linked}}}\right)\right] \times 100\% \tag{5.2}$$

where $(\mathrm{C/N})_{\text{observed}}$ is the C/N ratio measured by XPS. A fully cross-linked PA, $(\mathrm{C/N})_{\text{fully cross-linked}}$ has a C/N ratio of 6 and theoretically fully linear PA, $(\mathrm{C/N})_{\text{fully linear,}}$ has a C/N ratio of 7.5.

The change of elemental composition as a result of adding a novel monomer at different loadings during the IP process can also be quantitatively analyzed by XPS, as shown in Table 5.6. An et al. (2011) reported that the

TABLE 5.5

Comparison between Data Obtained from XPS and Calculated Data on TFC NF Membranes Prepared from IP of PIP and *mm*-BTEC

	Unit	Chemical Structure	C (%)	O (%)	N (%)	C/O	C/N	O/N
Possible outcomes	Linear structure with two pendant—COOH groups (A)	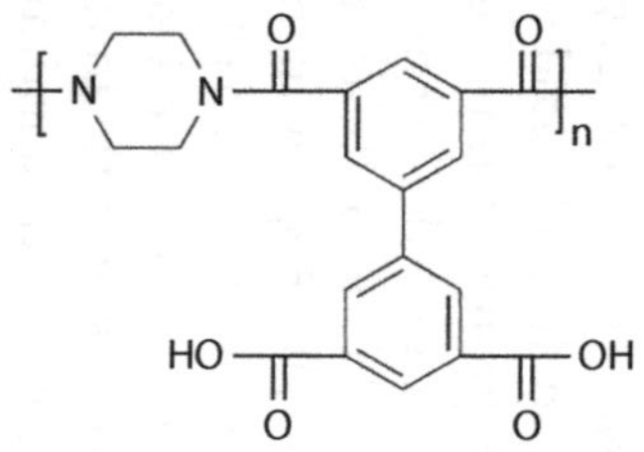	71.42	21.42	7.14	3.33	10.00	3.00
	Cross-linked with a pendant —COOH group (B)	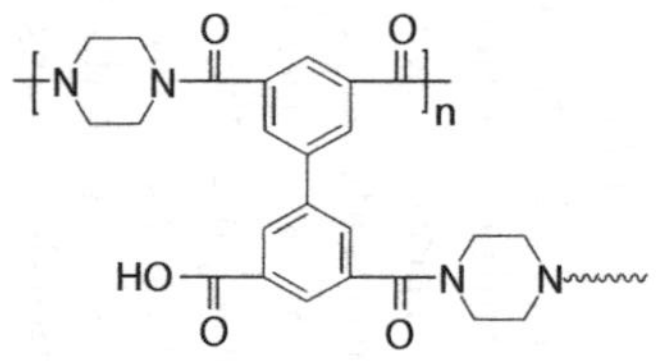	73.33	16.66	10.00	4.40	7.33	1.66
	Cross-linked with a pendant —NH– group (C)	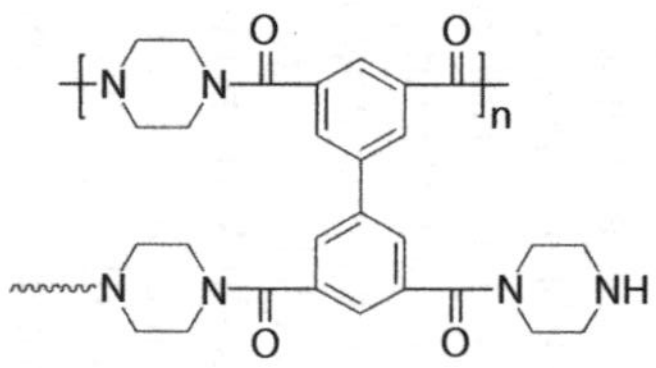	74.28	11.43	14.28	6.50	5.20	0.80

(Continued)

TABLE 5.5 (*Continued*)

Comparison between Data Obtained from XPS and Calculated Data on TFC NF Membranes Prepared from IP of PIP and *mm*-BTEC

Unit	Chemical Structure	C (%)	O (%)	N (%)	C/O	C/N	O/N
Totally cross-linked without pendant –COOH and –NH– groups (D)	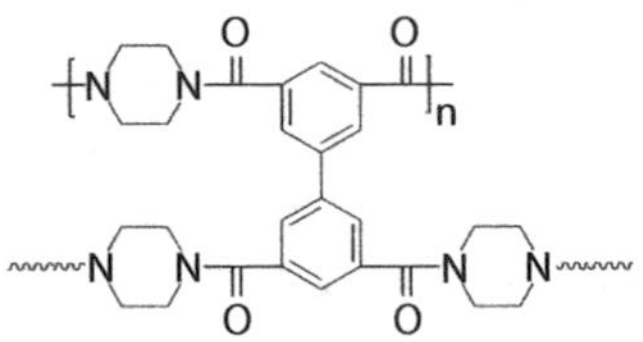	75.00	12.50	12.50	6.00	6.00	1.00
Linear structure with two pendant –NH– groups (E)	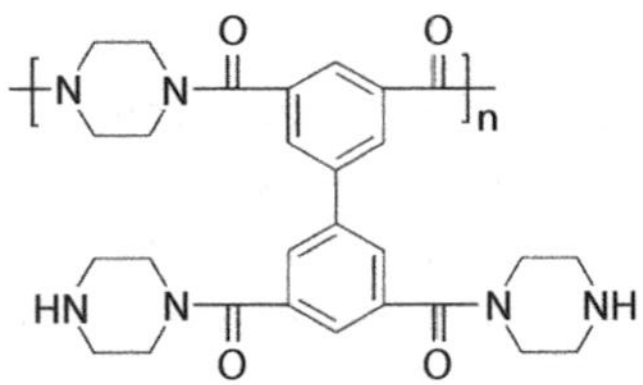	73.68	10.53	15.79	7.00	4.67	0.67
Results obtained from XPS		77.20	10.33	12.47	7.47	6.19	0.83

Source: Reprinted from *Journal of Membrane Science*, 378, Wang, H., Zhang, Q., Zhang, S., Positively charged nanofiltration membrane formed by interfacial polymerization of 3,3′,5,5′-biphenyl tetraacyl chloride and piperazine on a poly(acrylonitrile) (PAN) support, 243–249, Copyright 2011, with permission from Elsevier.

TABLE 5.6

Effect of Novel Monomer Loading on the Elemental Composition of TFC NF Membranes

Parameter	Loading	C (%)	O (%)	N (%)	C/O[a]	C/N[a]	O/N[a]	Remarks
Effect of PVA on the properties of PIP–TMC PA layer (An et al., 2011)	0% PVA[b]	72.37	13.99	13.63	5.17	5.31	1.03	High O/N ratio indicates the presence of more PVA in PA layer
	8% PVA	71.79	14.82	13.39	4.84	5.36	1.11	
	16% PVA	69.35	17.55	13.10	3.95	5.29	1.34	
Effect of DABA on the properties of MPD–TMC PA layer (Wang et al., 2010)	0 w/v% DABA[c]	76.24	12.16	11.61	6.27	6.57	1.05	High O/N ratio reveals the lower degree of CL process, owing to presence of free carboxylic acid groups
	0.05 w/v% DABA	74.08	13.96	11.96	5.31	6.19	1.17	
	0.15 w/v% DABA	73.76	15.61	10.63	4.73	6.94	1.47	
	0.25 w/v% DABA	75.53	15.34	9.12	4.92	8.28	1.68	
Effect of β-CD on the properties of TEOA–TMC PA layer (Wu et al., 2013b)	0 w/v% β-CD[d]	67.10	27.84	5.06	2.41	13.26	5.50	High ratio of O/N is due to large number of hydroxyl groups and/or lower CL PA structure
	1.8 w/v% β-CD	68.47	27.38	4.15	2.50	16.50	6.59	

[a] The elemental ratio is standardized for easy comparison.

[b] The percentage of PVA added was calculated based on the mass fraction of PVA to PIP. The concentrations of PIP and TMC during IP process were 0.35 and 0.2 wt/v%, respectively.

[c] The concentrations of MPD and TMC were 1.75–2.00 and 0.1 wt/v%, respectively.

[d] The concentrations of TEOA and TMC were 6 and 0.6 wt/v%, respectively.

O/N ratio of the PA layer increased from 1.03 to 1.34 when the mass fraction of polyvinyl alcohol (PVA) to PIP was increased from 0 to 16 w/v%. As PVA is a polymer with a significant number of hydroxyl groups, the increase in the composition of the O element is strongly linked to the presence of PVA in the PIP–TMC matrix. The introduction of a novel monomer—triamine 3,5-diamino-N-(4-aminophenyl) benzamide (DABA) into MPD–TMC PA layer was also found to increase the O/N ratio, owing to the higher content of free carboxylic acid groups (Wang et al., 2010). The increase in O/N ratio upon addition of DABA, however, has negative impact on the PA CL degree.

In the research work of Wu et al. (2013b), it was reported that the value of O/N increased remarkably from 5.50 to 6.59 when only 1.8 w/v% β-cyclodextrin (β-CD) was introduced into the TEOA–TMC PA membrane. The increase in O/N value might be due to the following reasons: (1) the large number of hydroxyl groups in β-CD and (2) lower CL structure caused by the introduction of β-CD. It is undeniable that there are many potential artifacts in an XPS experiment that make the interpretation of results difficult. These artifacts are often related to the sample preparation and/or sophisticated CL PA layer resulting from the use of more than two active monomers.

5.2.4 X-Ray Diffractometry

Recent advances in incorporating TFC membrane with nanomaterials such as CNT, zeolite, SiO_2, TiO_2, silver, etc. have afforded a viable composite membrane with overwhelming performance for liquid separation processes. To analyze the crystalline properties of composite membrane upon the addition of nanofillers on its thin film layer, an X-ray diffractometer with CuKα radiation ($\lambda = 1.54$ Å) can be employed. The X-ray diffraction (XRD) patterns are often recorded over a specified degree range with a scanning speed of 0.02–0.03° s^{-1} (2θ). The degree range employed during XRD analysis is strongly dependent on the diffraction data of nanoparticles added.

To really confirm the properties of nanoparticles, one can refer to the standard diffraction data for crystalline materials published by The International Centre for Diffraction Data (http://www.icdd.com). Several nanoparticles that have been analyzed by researchers with respect to XRD patterns for TFN membrane include NaA zeolite (peaks at 2θ of 12.5°, 16.1°, 21.7°, 24°, 27.1°, and 30°) (Jeong et al., 2007), NaX zeolite (2θ of 6°, 16°, and 27°) (Fathizadeh et al., 2011), MCM-41 SiO_2 nanoparticles (2θ of 2.1°, 3.8°, 4.4°, and 5.8°) (Yin et al., 2012), aluminosilicate SWNT (2θ of 2.8°, 8.8°, 13.4°, 26.5°, and 40°) (Barona et al., 2012), and TNTs (2θ of 10.8°, 25.2°, 29.9°, and 48.8°) (Emadzadeh et al., 2015). Besides identifying crystalline phase of nanoparticles, the data obtained from XRD can be used to determine the average size of nanoparticles using Scherrer's equation.

$$D = \frac{0.9\lambda}{\beta \cos\theta} \tag{5.3}$$

where D is the crystallite size, λ is the wavelength of the radiation ($\lambda = 1.5418$ Å), θ is Bragg's angle, and β is the full width at half maximum (radian).

It is not very common to use XRD to examine the changes in the surface crystallization of the TFN membrane. Up to date, there is still lack of XRD studies reporting the changes in membrane crystalline structure upon the addition of nanomaterials. It is mainly due to the relatively new TFN technology compared to the TFC membranes that are prepared without any nanomaterials. Nevertheless, Kim et al. in 2012 have successfully identified the presence of silver (Ag) particles in the top PA layer of the TFN membrane using an X-ray diffractometer. Based on the XRD data obtained (Figure 5.2), three diffraction peaks (37.5°, 45°, and 65°) belonging to the occurrence of Ag were found, confirming the existence of Ag within the PA matrix. Ghanbari et al. (2015) also confirmed the successful incorporation of HNTs in the PA layer based on the XRD pattern. However, by comparing the HNTs with the TFN membrane, it was found that one of the characteristic peaks of HNTs ($2\theta = 24.6°$) was shifted slightly to 23.2° as observed in the TFN membrane. This can be possibly due to the *in situ* polymerization phenomenon that has caused the intercalation of polymer chain into the intralamellar layer of the HNT and affected its characteristics.

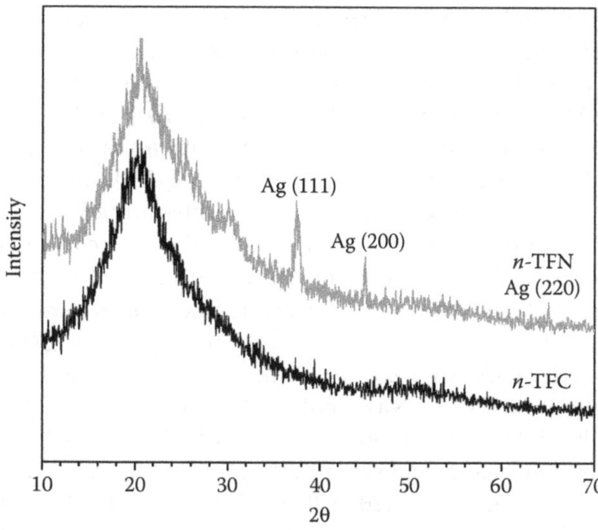

FIGURE 5.2
XRD patterns of n-TFC (PA: MPD–TMC, substrate: PSF incorporated with CNTs) and n-TFN membranes (PA: MPD–TMC–Ag, substrate: PSF incorporated with CNTs). (Reprinted from *Journal of Membrane Science*, 394–395, Kim, E. et al., Development of nanosilver and multi-walled carbon nanotubes thin-film nanocomposite membrane for enhanced water treatment, 37–48, Copyright 2012, with permission from Elsevier.)

5.2.5 Nuclear Magnetic Resonance Spectroscopy

Compared to the previously mentioned techniques, relatively few studies have been published on the correlation between composite membrane performance and its PA inherent properties using NMR spectroscopy. Since NMR is a technique to exploit the magnetic properties of atomic nuclei, it is thus useful in characterizing the organic structure of a newly synthesized monomer (Liu et al., 2012c; Xie et al., 2012; Wang et al., 2013) and evaluating any change in the PA cross-linked structure before and after chlorine exposure (Glater et al., 1994; La et al., 2010). The advantage of NMR spectroscopy in analyzing membrane material is due to the low frequency (tens of kHz) molecular motion of solid polymers (Kwak and Woo, 1999), causing the use of spin–lattice relaxation times of proton in the rotating frame (T_{1p}) to be very useful. Another unique advantage of this technique includes its ability to study the molecular structure of a sample in both solution and solid state.

For solution-state NMR analysis, the polymeric membrane sample is dissolved in dimethylsulfoxide-d_6 solvent to form a 10% (w/v) solution while for solid-state analysis, membrane samples with a mass of 50–60 mg are cut into small pieces and packed into magic-angle spinning rotors for stable spinning (Lee et al., 2011). Generally, high-resolution solid-state NMR spectra can provide the same type of information as shown in solution NMR spectra, provided it is equipped with techniques/equipment such as magic-angle spinning, cross polarization, special pulse sequences, enhanced probe electronics, etc. Nevertheless, the main advantage of using solid-state NMR spectroscopy is its capability to directly handle polymeric membrane (with no dissolution) and provide structural information nondestructively.

Figure 5.3 presents the NMR spectra of newly developed chlorine-resistant TFC membrane and typical MPD–TMC membrane after being exposed to 500 ppm hypochlorous acid (HOCl) solution at pH 5.5 for 17 h (La et al., 2010). Based on the NMR spectral analysis, there were significant new peaks found in the typical MPD–TMC membrane upon chlorine exposure compared to insignificant change in the peaks found in the newly developed membrane made of HFA and TMC. The changes in the peaks (MPD–TMC membrane) are consistent with the chlorination mechanism resulting from *N*-chlorination followed by migration of Cl-substituent onto the benzene ring. Instead of using NMR to evaluate membrane chlorine resistance, many prefer to conduct a filtration test to assess the changes in membrane performances with respect to flux and salt rejection upon chlorine exposure. Nevertheless, it must be pointed out that NMR is a qualitative approach that can provide in-depth description on the membrane property, supporting quantitative results obtained from conventional filtration experiments. Furthermore, it can complement other structural and analytical techniques such as XPS.

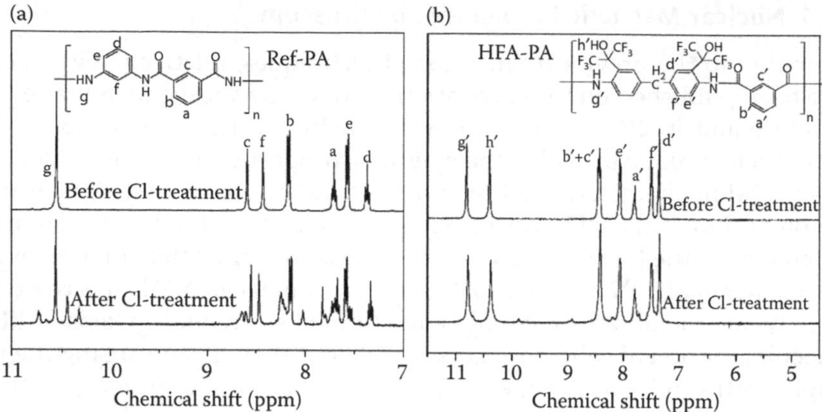

FIGURE 5.3
^{1}H NMR spectra of TFC membrane before and after chlorine treatment. (a) Ref-PA membrane (made of MPD and TMC) and (b) HFA-PA membrane (made of (HFA) containing aromatic diamine and TMC). (La, Y. et al. 2010. Novel thin film composite membrane containing ionizable hydrophobes: pH-dependent reverse osmosis behavior and improved chlorine resistance. *Journal of Materials Chemistry*. 20, 4615–4620. Reproduced by permission of The Royal Society of Chemistry.)

5.3 Methods for Physical Properties Assessment

5.3.1 Electron Microscopy

The electron microscopic tools are of great importance for investigating composite membrane morphology at a length scale spanning from a few nanometers to hundreds of micrometers. Three types of electron microscopy techniques that are commonly used to study composite NF membrane morphological properties are SEM, FESEM, and TEM. The main advantage of these techniques is that the visual information of the membrane morphology can be obtained directly at the desired resolution. To decide which technique is suitable for a sample, one needs to know the maximum level of resolution required for the sample. It is because different microscopy techniques offer different resolutions, depending on the working principle.

To prepare a membrane sample for SEM/FESEM analysis, the backing fabric (or nonwoven fabric) of the TFC/TFN membrane needs to be gently peeled off before immersing into liquid nitrogen. This is because it is very difficult for the composite membrane together with the backing fabric to be fractured even in liquid nitrogen, and may risk damaging the membrane structure. The specimen for cross sectional analysis is prepared by cryogenically fracturing after immersing the sample into liquid nitrogen for a few minutes. The sample is then carefully mounted onto a stub before it is sputter-coated under vacuum with an ultrathin layer made of either gold (Au) or platinum (Pt) as a conducting material.

During sample analysis, employing excessive high electron-beam energy is not encouraged as it could cause damage to the membrane sample. Energy ranging of 10–20 kV is the typically applied condition for a polymeric membrane sample. As SEM/FESEM requires the complete dehydration of the membrane sample during analysis, the drying process might change the structure of membranes especially those membranes made from hydrophilic polymers. Sample integrity is somewhat compromised by artifacts. Furthermore, the coating process prior to sample analysis might result in size diminution on the membrane surface.

Figure 5.4 compares the surface morphologies of TFC membranes analyzed by SEM and FESEM under different magnifications together with their respective preparation conditions (Ghosh and Hoek, 2009; Fathizadeh et al., 2011). The "ridge-and-valley" (also named as "leaf-like") morphology of a TFC membrane shown in Figure 5.4a and c is the well-known characteristic of an interfacially polymerized PA membrane. As mentioned earlier, the advantage of using a microscope is to obtain direct information concerning the membrane structure. Figure 5.4b and d on the other hand shows the changes in the PA structure by using different substrate properties and with the addition of nanoparticles into hexane (during IP process), respectively. It was found that the enhanced hydrophilic nature of the PSF substrate upon the addition of hydrophilic additive could form a "nodular" film (Ghosh and Hoek, 2009) while the addition of hydrophilic NaX zeolite into the PA matrix could result in the formation of some spherical and cylindrical shapes on the PA layer (Fathizadeh et al., 2011).

Besides examining the surface morphology of a composite membrane, SEM/FESEM can also be used in examining the cross sectional structure of the PA layer as well as the microporous substrate. Typically, the thin PA layer on the top of the substrate layer can be easily distinguished when visualized under magnification of around 20,000×. Nevertheless, the irregular morphology of the PA layer would preclude quantification of a single film layer thickness. It is very often reported that PA thin films have thickness between 100 and 500 nm.

In order to obtain high-resolution images (with magnification up to 100,000×) on NF membrane morphology with sufficient contrast, TEM is highly recommended. However, the sample preparation for TEM is relatively difficult and complex compared to that of SEM/FESEM. The membrane sample needs to be first embedded with epoxy resin before being cut in a Leica ultramicrotome under cryogenic condition. Other significant drawbacks of TEM include large instrument size, high cost, special training required for operation and analysis, potential artifacts from sample preparation, and small viewing size. Compared to the energy applied in SEM/FESEM, the energy required for TEM analysis is significantly higher, that is, between 80 and 100 kV.

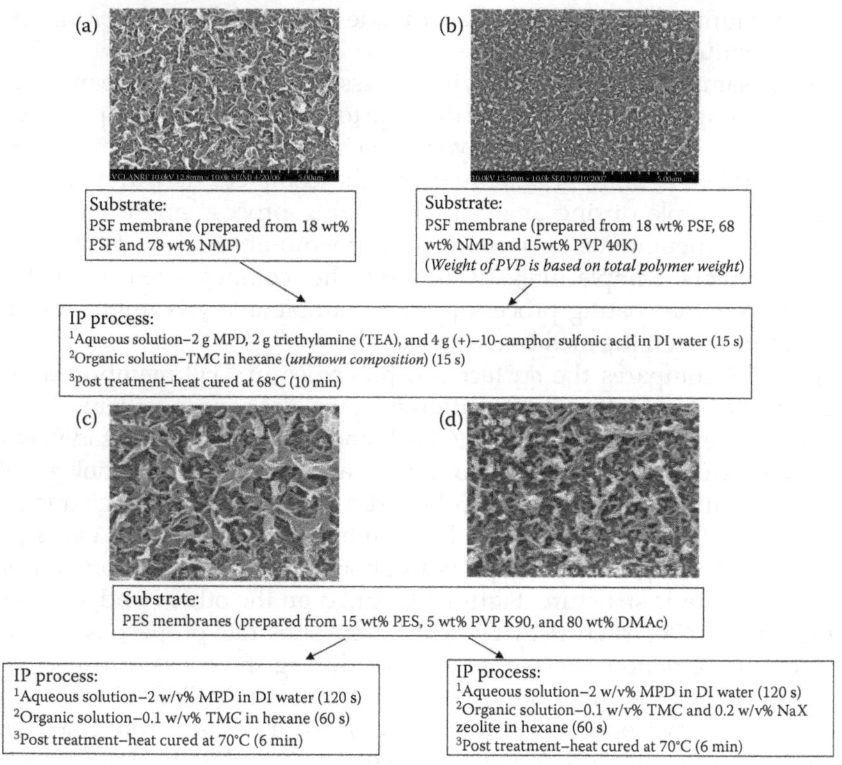

FIGURE 5.4
Surface morphologies and preparation conditions of TFC PA membranes. (a), (b) SEM images captured at magnification of 10,000×. (Reprinted from *Journal of Membrane Science*, 336, Ghosh, A.K., Hoek, E.M.V., Impacts of support membrane structure and chemistry on polyamide–polysulfone interfacial composite membranes, 140–148, Copyright 2009, with permission from Elsevier.) (c), (d) FESEM images captured at magnification of 15,000×. (Reprinted from *Journal of Membrane Science*, 375, Fathizadeh, M., Aroujalian, A., Raisi, A., Effect of added NaX nano-zeolite into polyamide as a top thin layer of membrane on water flux and salt rejection in a reverse osmosis process, 88–95, Copyright 2011, with permission from Elsevier.) The superscript number in the IP process indicates the sequence of process while the time (in bracket) shows the immersion/process period.

Figure 5.5 shows the TEM images of a novel PA layer (incorporated with nanofillers) deposited on the PSF substrate (Jeong et al., 2007; Kim et al., 2012; Yin et al., 2012). It is clearly shown that only with the use of TEM, the presence of nanofillers (darker spots appeared) within/on the PA layer could be clearly seen. Moreover, the aggregation of MCM-41 nanoparticles could also be detected on the top surface of the thin film layer (Figure 5.5b), showing TEM could provide more information on the morphology particularly on a nanoscale level compared with SEM/FESEM. Nevertheless, the main problem of using TEM for sample analysis is its extremely small region analyzed, which causes the resulting images may not be indicative of the properties

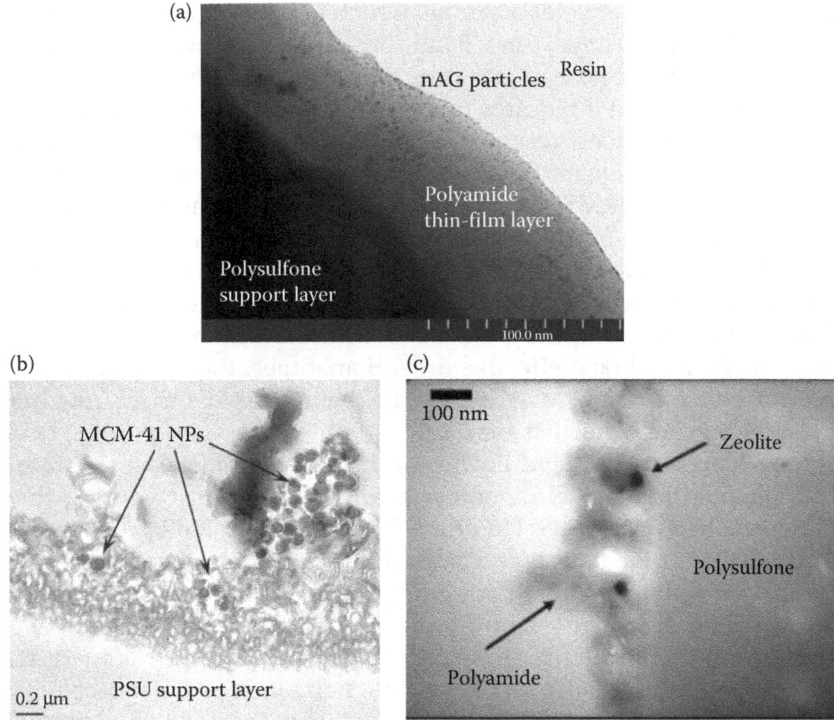

FIGURE 5.5
TEM images of the cross-section of TFN membrane incorporated with (a) silver (Ag) nanoparticles (Reprinted from *Journal of Membrane Science*, 394–395, Kim, E. et al., Development of nanosilver and multi-walled carbon nanotubes thin-film nanocomposite membrane for enhanced water treatment, 37–48, Copyright 2012, with permission from Elsevier), (b) MCM-41 silica nanoparticles (Reprinted from *Journal of Membrane Science*, 423–424, Yin, J. et al., Fabrication of a novel thin-film nanocomposite (TFN) membrane containing MCM-41 silica nanoparticles (NPs) for water purification, 238–246, Copyright 2012, with permission from Elsevier), and (c) NaA zeolite nanoparticles (Reprinted from *Journal of Membrane Science*, 294, Jeong, B.-H. et al., Interfacial polymerization of thin film nanocomposites: A new concept for reverse osmosis membranes, 1–7, Copyright 2007, with permission from Elsevier).

of the entire membrane surface. In order to obtain complementary information on membrane properties, analyses using other types of instruments are required, in addition to conventional filtration experiments.

5.3.2 Atomic Force Microscopy

As a complement to electron microscopies, atomic force microscopy (AFM) is employed for the morphological characterization of the NF membrane surface. A cantilever with a sharp tip at its end having a radius of less than 10 nm is used to scan the surface of a membrane with a constant force of 5 N m^{-1}.

The typical scanning area for a sample which is fixed on a specimen holder used to be 10 μm × 10 μm, 5 μm × 5 μm, or 2 μm × 2 μm with scanning rate set at 1 Hz. In certain cases, a scanning area as large as 50 μm × 50 μm can also be performed, but it requires more time to complete one sample analysis. The surface roughness can be quantified with respect to average roughness (R_a), root mean square (RMS) roughness, and surface area difference. Of these parameters, RMS roughness, which is defined as the mean of the root for the deviation from the standard surface to the indicated surface, is the key physical parameter reported by researchers to reveal the extent and occurrence of composite membrane surface roughness. Generally speaking, the higher the RMS value, the greater the membrane surface roughness and the higher the membrane effective area. Sometimes, the increase in membrane water flux is partly associated with the increased surface roughness that provides greater effective area.

Figure 5.6 presents the AFM images of the changes of PA layer surface by adding nanoparticles/novel monomer into the PA layer or varying the IP conditions (Chen et al., 2008; Hu et al., 2012; Yin et al., 2012; Misdan et al., 2015). The color of the images indicates the vertical deviations of the membrane surface with the light regions being the peaks and dark regions being the depressions. As can be seen from the images, by varying the parameters during the PA layer synthesis process, the changes on the membrane surface roughness could be visually seen; even if it happened on a nanoscale level. The increase in membrane surface roughness upon the addition of nanoparticles (Figure 5.6a and b) could be due to the aggregation of nanoparticles on the surface of the thin film layer or enlargement of the effective membrane surface area. The decrease in surface roughness (Figure 5.6c) on the other hand is likely caused by the lower CL reaction rate and/or stronger intermolecular hydrogen bonding. With respect to reaction time, it is found that longer polymerization time tends to increase PA surface roughness (Figure 5.6d) as adjacent granule-like structures tend to cross-link and form a supergranule-like morphology.

Although increased surface roughness could contribute to increasing water flux, it could also possibly increase the fouling tendency by increasing the rate of foulant attachment onto the membrane surface (Ng et al., 2013). Typically, an NF membrane with a surface comprising of ridges and valleys can be considered smooth if its RMS roughness is of about 30–50 nm.

5.3.3 Positron Annihilation Spectroscopy

To provide more complete information on the same composite membranes that have been characterized by conventional techniques such as FESEM and AFM, a relatively new approach—positron annihilation spectroscopy (PAS) is recommended as an advanced analytical tool to detect molecular vacancies and pores (in nm) present in membrane materials. As PAS is a nondestructive and highly informative technique for characterizing dense

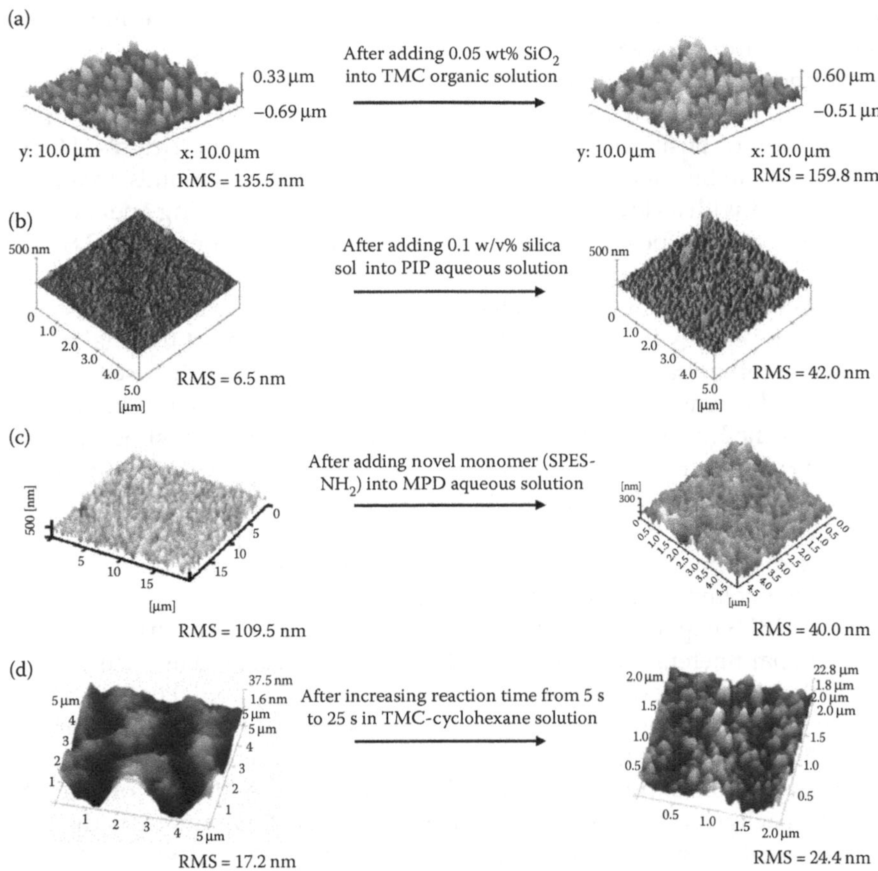

FIGURE 5.6
AFM images of the changes of surface morphologies, (a) MPD–TMC membrane before and after addition of MCM-41 silica with particle size of about 100 nm (Reprinted from *Journal of Membrane Science*, 423–424, Yin, J. et al., Fabrication of a novel thin-film nanocomposite (TFN) membrane containing MCM-41 silica nanoparticles (NPs) for water purification, 238–246, Copyright 2012, with permission from Elsevier), (b) PIP–TMC membranes before and after addition of SiO_2 with particle size of 11 nm (Reprinted from *Desalination*, 301, Hu, D., Xu, Z.-L., Chen, C., Polypiperazine-amide nanofiltration membrane containing silica nanoparticles prepared by interfacial polymerization, 75–81, Copyright 2012, with permission from Elsevier), (c) MPD–TMC membrane before and after addition of SPES-NH_2 (Reprinted from *Journal of Membrane Science*, 310, Chen, G. et al., Novel thin-film composite membranes with improved water flux from sulfonated cardo poly(arylene ether sulfone) bearing pendant amino groups, 102–109, Copyright 2008, with permission from Elsevier), and (d) PIP–TMC membrane prepared at different reaction times (Reprinted with permission from Misdan, N., Lau, W.J., Ismail, A.F. 2015. Physicochemical characteristics of poly(piperazine-amide) TFC nanofiltration membrane prepared at various reaction times and its relation to the performance. *Journal of Polymer Engineering.* 35(1), 71–78. Copyright 2015 Walter de Gruyter GmbH.)

polymeric membranes, it has received a lot of attention in the membrane community in recent years.

In principle, the characterization of membrane free-volume hole properties using PAS is based on the detection of γ-radiation produced during the annihilation of positrons inside the probe materials. Properties of TFC NF membrane can be characterized by two different PAS techniques, that is, (a) PAS coupled with a slow positron beam of doppler broadening energy spectra (DBES) and (b) positron annihilation lifetime spectroscopy (PALS). For both techniques, a ^{22}Na radioisotope source is used to generate a positron beam. When positrons are injected into polymeric membranes, they lose their kinetic energy in several ways due to the occurrence of thermalization in the membrane sample. Theoretically, the positron either diffuses into the media and becomes annihilated as a free positron, or combines with an electron from the material to form a bound-state positronium (Ps) of two spin states (*para*-Ps (*p*-Ps) and *ortho*-Ps (*o*-Ps)) or loses energy due to interaction with electrons (Kim, 2005; Tung et al., 2009). As *o*-Ps has a reasonable long lifetime of 142 ns in vacuum compared to *p*-Ps and free positron which have lifetimes of 0.125 and 0.45 ns, respectively, it has become the choice for measuring free-volume hole properties of membrane.

In the DBES experiments, the spectra obtained are characterized by two different parameters (S and R parameter) that are measured as a function of incident positron energy at room temperature ranging from 0 to 30 keV. The typical total count for each DBES spectrum is one million with a counting rate of 2000–2800 counts per second. The S parameter is defined as the ratio of integrated concentration between 510.3 and 511.7 keV to the total counts after the background is properly deducted while the R parameter is the ratio of total counts from the valley region with energy width between 364.2 and 496.2 keV (from 3γ annihilation) to total counts from the 511 keV peak region with energy width between 504.35 and 517.65 keV (from 2γ annihilation) (Tung et al., 2009). In brief, a large value of the S parameter indicates a large free-volume depth profile while a large R parameter value corresponds to the existence of large cavities.

Figure 5.7 shows the values of S and R parameters obtained as a function of the incident positron energy for a commercial NF membrane (ESNA1-K1, Nitto-Denko) (Nanda et al., 2011). The mean depth shown on the upper x-axis is determined from incident positron energy using the equation as follows:

$$Z(E_+) = \left(\frac{40 \times 10^{-3}}{\rho}\right) E_+^{1.6} \tag{5.4}$$

where E_+ is the incident positron energy, Z is the mean depth, and ρ is the density of polymeric material. From the figure, the low values of S parameter and high values of R parameter near the PA layer surface are mainly due to the back diffusion of Ps at the membrane surface in the range of 1–20 nm

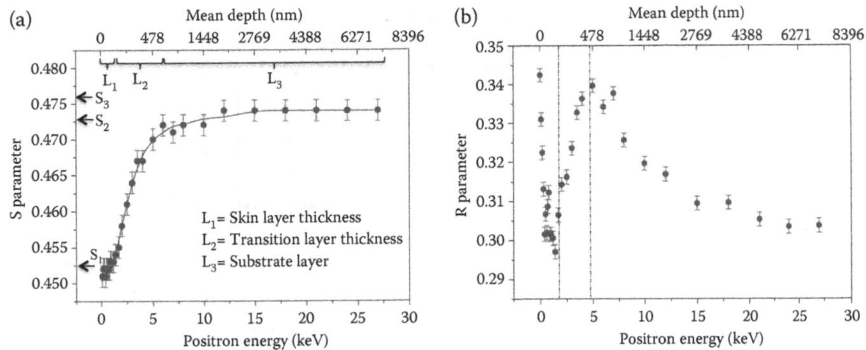

FIGURE 5.7
Parameter as a function of incident positron energy and mean depth of commercial ESNA1-K1 NF membrane (Nitto-Denko), (a) *S* parameter and (b) *R* parameter. (Reprinted from *Journal of Membrane Science*, 343, Tung, K.-L. et al., Characterization of multilayer nanofiltration membranes using positron annihilation spectroscopy, 147–156, Copyright 2009, with permission from Elsevier.)

(Tung et al., 2009; Ong and Chung, 2012). This phenomenon is commonly found in PAS studies.

Using the VEPFIT program, the PAS data of the S parameter revealed that the composite membrane consisted of three layers: a PA skin layer (S_1), a transition layer (S_2), and a substrate layer (S_3). Due to the penetration limitation of positron (~8 μm) used in PAS studies, determination of total substrate layer thickness is generally ignored. Nevertheless, the thickness of the PA skin layer (L_1) and transition (L_2) layer can be quantitatively determined as shown in Figure 5.7a. With respect to the R parameter, the minimum value as found near the membrane surface indicates that the PA layer is the densest layer among the three different layers observed in the NF membrane. Meanwhile, the plateau of the R parameter found in the transition layer indicates the formation of a loose structure (high porosity) between the PA layer and substrate layer.

In addition to the *S* and *R* parameters, mean radius and cavity volume of a membrane can also be determined by PAS, but only through PALS technique. In this experiment, each spectrum is counted for two million at a counting rate of 200–300 counts per second. The obtained PAS data can be then interpreted by fitting the data into three lifetimes using the PATFIT program and into continuous lifetime distribution using the MELT program. The pick-off lifetime of *o*-Ps (τ_p) has been proven to be highly correlated with the mean radius of the cavity (R), assuming that the Ps is localized in a spherical potential well with an electron layer thickness (ΔR) equal to 1.66 Å.

$$\tau_p = \frac{1}{2}\left[1 - \frac{R}{R+\Delta R} + \frac{1}{2\pi}\sin\left(\frac{2\pi R}{R+\Delta R}\right)\right]^{-1} \tag{5.5}$$

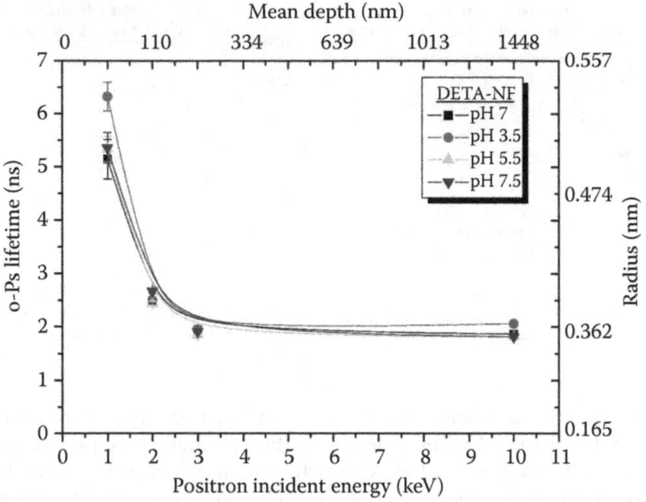

FIGURE 5.8
o-Ps lifetime and mean radius of cavity as a function of positron incident energy for an NF membrane prepared from IP of DETA and TMC over a PAN support membrane. (Reprinted from *Journal of Membrane Science*, 382, Nanda, D. et al., Characterization of fouled nanofiltration membranes using positron annihilation spectroscopy, 124–134, Copyright 2011, with permission from Elsevier.)

Figure 5.8 shows the changes in the lifetime of *o*-Ps and mean radius of an NF membrane as a function of the positron incident energy after the membrane was tested using calcium sulfate feed solutions at different pH environments (Nanda et al., 2011). The results revealed that the cavity size of the NF membrane remained almost the same regardless of the feed solution pH, indicating the properties of the membrane were not affected by the solution pH ranging between 3.5 and 7.5. It must be noted that the longer lifetime of *o*-Ps near the membrane top surface can be explained by the back diffusion of *o*-Ps from the membrane surface to the vacuum. In order to further calculate the volume of cavity (V_c) of NF membrane, Equation 5.6 is employed by inserting the mean radius obtained from Equation 5.5.

$$V_c = \frac{4}{3}\pi R^3 \tag{5.6}$$

Since FESEM and AFM are only useful for detecting static defects near the PA surface, the use of PAS in membrane characterization should not be treated as a competitive but as a complementary technique to overcome the shortcomings of conventional techniques, providing insight into the interpretation of the cavity properties of the membrane at various depths.

5.3.4 Surface Contact Angle

Wettability plays a very crucial role in determining water permeability of a membrane. It is generally known that the higher the degree of wettability, the greater the amount of water (membrane flux) can be produced and vice versa. In addition to good water productivity, a membrane of high wettability degree is advantageous to be operated at a lower pressure which reduces energy consumption. To evaluate membrane wettability, two parameters are considered, (a) contact angle which is determined using contact angle analyzer and (b) water uptake which is assessed by measuring the weight of the water-saturated membrane sample and the dry weight of the same sample. In this subsection, only the details regarding the membrane contact angle will be discussed.

In assessing the membrane contact angle, one needs to understand the phenomenon of the nature of water molecules on a clean membrane surface. When the force of adhesion (i.e., force with which water molecules adhere to a membrane surface) is greater than the cohesion force of water molecules, the water molecules tend to wet the surface, resulting in lower value of contact angle as shown in Figure 5.9. An ideal superhydrophilic membrane surface tends to display a zero contact angle, but it has never been achieved so far. In PA, the main components affecting hydrophilicity are amide bond, amine end group, and carboxylic acid end group. Comparatively, the amine end and carboxylic acid groups are more hydrophilic than the amide bonds of the PA chain (Li et al., 2008).

Contact angle measurement is typically performed at room temperature by a contact angle goniometer. Simply making a water drop followed by measuring it provides a result referred to as static contact angle measurement. For the dynamic measurement, drops can be made to have advanced edges by the addition of liquid while receded edges can be produced by allowing sufficient evaporation or by withdrawing liquid from the drop. Alternately, both advanced and receded edges are produced when the stage on which the membrane is held is tilted to the point of incipient motion. Briefly, for a membrane with a relatively smooth and homogeneous surface, the general practice for contact angle measurement is based on the static

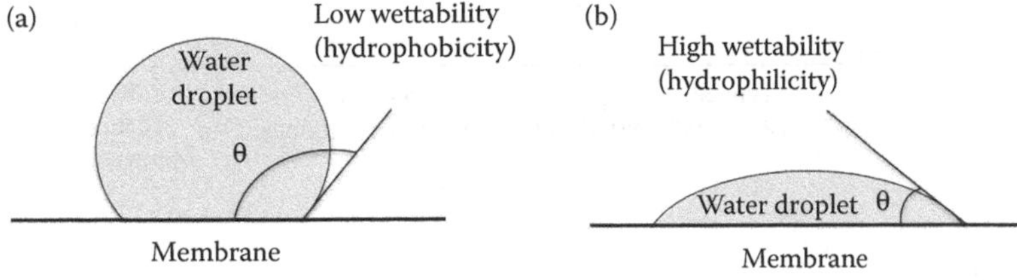

FIGURE 5.9
Schematic diagram of a single water droplet on membrane surface, (a) hydrophobic surface ($\alpha > 90°$) and (b) hydrophilic surface ($\alpha < 90°$).

one. Although the captive bubble method can also be to be used, one must be aware of the possibility of the membrane swelling after immersion into a liquid.

To perform a static contact angle measurement, a pure water droplet of 2–5 mg (or μL) is dropped carefully onto a dry membrane surface using a motor-driven micro-syringe. In order to produce high accuracy measurement, replication of 8–12 times on different locations of the same membrane sample is highly recommended to yield a reliable result. Since the contact angle on the membrane surface will decrease with time due to one or more of the following reasons—evaporation, relaxation of the liquid, chemical reaction between the solid and the liquid, and/or liquid being absorbed into the solid, it is of interest to take the measurement immediately after the water is contacted on the membrane surface. The Wenzel and Cassie effect should also be taken into consideration during measurement as membrane surface roughness could have an impact on the surface wettability and thus on the contact angle. The Wenzel equation is used by assuming that the liquid penetrates into the roughness grooves while the Cassie equation assumes that the liquid does not penetrate into the grooves.

Figure 5.10 presents the range of contact angle shown by different chemistry properties of the PA layer. However, due to overlapping of some data, the total number of data collected is more than what are seen in this figure. It must be pointed out that even if the PA layer is interfacially polymerized between the common MPD and TMC monomer, the exact contact angle of this particular PA layer is still hard to confirm as it can be influenced by many other parameters during the IP process. These include monomer

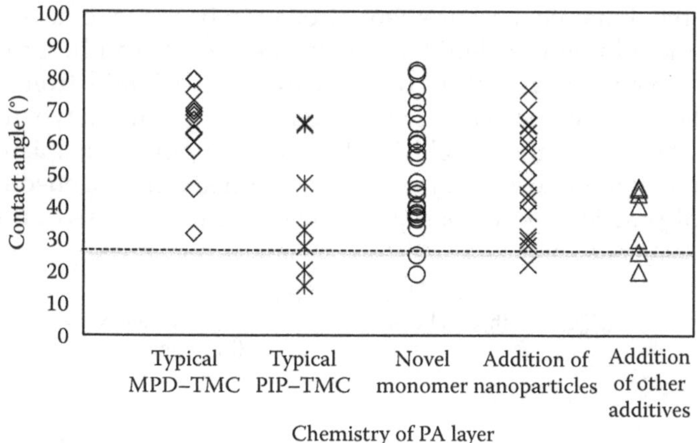

FIGURE 5.10
Range of contact angle shown by PA layers prepared from different recipes. (Reprinted with permission from Lau, W.J. et al., 2015. Characterization methods of thin film composite nanofiltration membranes: A review. *Separation and Purification Reviews*. 44, 135–156. Copyright 2015 Taylor & Francis.)

concentration, reaction time, type of organic solution, post-treatment condition, etc. Nevertheless, these data indicated that the decrease in the contact angle of typical MPD–TMC PA layer is still achievable with the use of a novel monomer and/or addition of hydrophilic nanofillers. For instance, the incorporation of these advanced materials in the PA matrix has shown potential to reduce the contact angle of PA to less than 25° which is hard to be achieved by a typical MPD–TMC PA layer.

5.4 Characterization on Membrane Permeability and Selectivity

5.4.1 Permeability

Before going into the details of the proper characterization methods used to determine membrane water flux and solute rejection, an attempt is made to show the relationship between NF permeability and salt rejection. Figure 5.11 summarizes the performances of NF membranes reported in more than 100 research articles published in peer-reviewed journals. This figure is the first analysis between water flux and salt separation efficiency of NF membranes ever presented. However, the use of different salt concentrations in the feed solution as well as different operating pressures applied by different researchers is the major obstacle in preparing the figure. In order

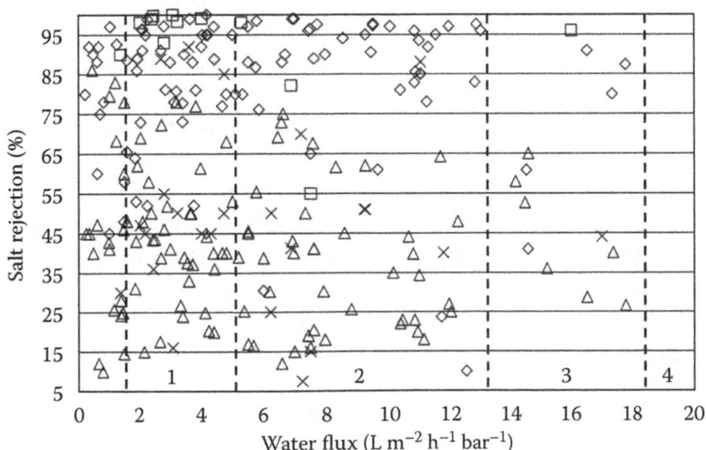

FIGURE 5.11
Salt rejection and water flux achieved by NF membranes published in the literature (◇—Na_2SO_4 rejection of synthesized NF, Δ—NaCl rejection of synthesized NF, □—Na_2SO_4 rejection of commercial NF, and ×—NaCl rejection of commercial NF).

to overcome the problem, only feed solutions containing low salt content (500–2000 mg L^{-1}) are considered, by assuming the minimal effect of osmotic pressure created by rejected ions on the water flux.

For the sake of simplicity and interpretation of the results reported in the literature, the water permeability coefficient is standardized to L m^{-2} h^{-1} bar^{-1}. As can be seen from the figure, the performance of NF membranes can be generally categorized into four regions in accordance to the water flux coefficient achieved. Mostly, the in-house made NF membranes are positioned in region 1 (with water flux of 1.5–5 L m^{-2} h^{-1} bar^{-1}) together with most of the commercial NF membranes available in the market. With the development of novel materials such as active monomers and nanoparticles, it is found that the water flux of in-house made NF membranes could be further improved to 5–13.5 L m^{-2} h^{-1} bar^{-1} as shown in region 2. However, empirical evidence (region 3) suggests that it is difficult to further increase the water flux of NF membrane without sacrificing selectivity. To develop superior NF membrane with high flux/high selectivity that falls in region 4 will necessitate a major paradigm shift, as it will require a membrane that does not follow the size exclusion/charge repulsion and solution-diffusion mechanisms.

Membrane water permeability can be evaluated through a simple filtration process. Due to the simplicity in terms of experimental setup and operation, the dead-end filtration mode is preferable for use during lab-scale experiments. Prior to the flux measurement, the NF membrane is subjected to a compaction process by filtering pure water (either deionized water or distilled water) at hydrostatic pressure 10–20% higher than the maximum operating pressure in order to avoid any further change in its structure during experiments. For instance, it is required to apply 11–12 bars for membrane compaction if the maximum operating pressure of filtration process is set at 10 bars. The typical period needed to achieve stable water flux varies depending on the NF membrane properties, but is often reported in the range of 30 min to several hours.

After achieving steady-state condition, the PWF of the NF membrane at a specific operating pressure can be evaluated. Besides PWF, another term that is used in the literature to name membrane water flux is pure water permeability (PWP). For clarity, the term PWF is being used exclusively in this book. When selecting water as feed solution, the main criterion is that it must have conductivity of less than 5 μS cm^{-1} to avoid any flux reduction due to osmotic pressure effect. The PWF of NF membrane tested at specific operating pressure can be determined by the expression as follows:

$$J_{PWF} = \frac{Q}{At} \tag{5.7}$$

where Q is the quantity of permeate, A is the effective membrane area, and t is the time required to obtain the quantity of Q. It must be pointed out that many different units are used in the literature to report the PWF of NF

TABLE 5.7

Conversion Table for Permeate Fluxes

Unit A[a] \ Unit B	$m^3\ m^{-2}\ s^{-1}$	$cm^3\ cm^{-2}\ h^{-1}$	gal ft^{-2} day^{-1}	L m^{-2} h^{-1}	L m^{-2} day^{-1}
$m^3\ m^{-2}\ s^{-1}$		3.6×10^5	2.1×10^6	3.6×10^6	8.6×10^7
$cm^3\ cm^{-2}\ h^{-1}$	2.8×10^{-6}		5.9	10	240
gal ft^{-2} day^{-1}	4.7×10^{-7}	1.7×10^{-1}		1.7	41
L m^{-2} h^{-1}	2.8×10^{-7}	0.1	0.59		24
L m^{-2} day^{-1}	1.2×10^{-8}	4.2×10^{-3}	2.5×10^{-2}	4.2×10^{-2}	

[a] The conversion table is prepared by converting value of Unit A to value of Unit B.

membranes. These include L m^{-2} h^{-1}, gal ft^{-2} day^{-1}, kg m^{-2} h^{-1}, m s^{-1}, m^3 m^{-2} h^{-1}, cm^3 cm^{-2} s^{-1}, etc. The respective conversion factors are given in Table 5.7. However, the unit L m^{-2} h^{-1} is used throughout this book for easy comparison. Furthermore, as there exists no standard method for the determination of PWF of NF membrane, one might find it rather difficult to evaluate the membrane performance reported by different researchers in which the membranes are tested at different operating pressures. The use of a standard unit—PWF coefficient (L_p) is, therefore, highly recommended. The value of L_p can be experimentally determined by plotting J_{PWF} against applied pressure with the assumption of null value of osmotic pressure, π. Typically, L_p of an NF membrane can vary from about 1 to 20 L m^{-2} h^{-1} bar^{-1}.

5.4.2 Selectivity

The retention of an NF membrane against a specific solute is governed not only by the sieving (size exclusion) effect but also the Donnan exclusion (charge repulsion) effect. Owing to these unique solute transport mechanisms, the separation performance of an NF membrane is generally assessed using two different types of test solutes, that is, neutral solutes which have molecular weight (MW) between 100 and 1000 Dalton (Da) (equivalent to g mol^{-1}) and charged solutes which have different charge numbers. The use of dye components is also reported for membrane rejection, but it is mainly used in characterizing SRNF membrane. More details regarding this characterization will be discussed in the following sections.

5.4.2.1 Neutral Solutes

Retention measurement of NF membrane using neutral solute is of importance for determining membrane MWCO. It must be noted that only neutral solutes are considered in MWCO determination as any charged solutes might influence the separation rate of NF membrane that generally possesses negative surface charge at neutral aqueous environment. Since pore dimensions on the NF membrane are in the near- and sub-nanometer range, selection

TABLE 5.8

Molecular Size and Stokes Radii of Typical Components Used as Markers in the Characterization of NF Membrane

Solute	MW (g mol^{-1})	r_s ($\times 10^{-9}$ m)
PEG 200	200	0.376
PEG 400	400	0.518
PEG 600	600	0.624
PEG 1000	1000	0.790
PEG 2000	2000	1.154
Glycerol	92	0.260
Glucose	180	0.365
Saccharose	342	0.471
Raffinose	504	0.584

of the size of neutral solutes plays a crucial role. It is highly recommended to use neutral solutes with MW that fall within the NF category as shown in Table 5.8. Commonly, these neutral solutes are selected from sugar-based solutes and/or polymers having the same structural unit such as PEG.

In order to minimize the effect of osmotic pressure due to the solute rejected, a feed solution containing low concentration of neutral solute (between 200 and 250 mg L^{-1}) is highly recommended for MWCO characterization. Furthermore, one must be always aware of the concentration polarization effect that could affect the accuracy of the true membrane rejection measurement. The concentration polarization modulus during membrane process can be expressed as

$$\frac{C_{io}}{C_{ib}} = \frac{\exp(J_v\delta/D_i)}{1+E_0[\exp(J_v\delta/D_i)-1]} \tag{5.8}$$

where C_{io} is the solute concentration adjacent to membrane surface, C_{ib} is the bulk solution concentration, J_v is the permeate volume flux, δ is the thickness of boundary layer, D_i is the diffusion coefficient of the solute, and E_o is the ratio of permeate solute concentration to solute concentration adjacent to membrane surface (C_{ip}/C_{io}). Ideally, no concentration polarization occurs when the modulus is 1.0. However, as the modulus deviates farther from 1.0 (i.e., $C_{io} > C_{ib}$), the impact of concentration polarization on membrane performance becomes increasingly significant. To minimize the impact of concentration polarization, the boundary layer thickness and filtrate flux need to be reduced. This can be practically achieved by applying high tangential flow velocity at the membrane surface and by operating the membrane at lower pressure.

The MWCO of a membrane is defined by the MW of the component that is retained by greater than 90%. Figure 5.12 shows the rejection profile of a commercial NF70 membrane as a function of MW of neutral solutes in

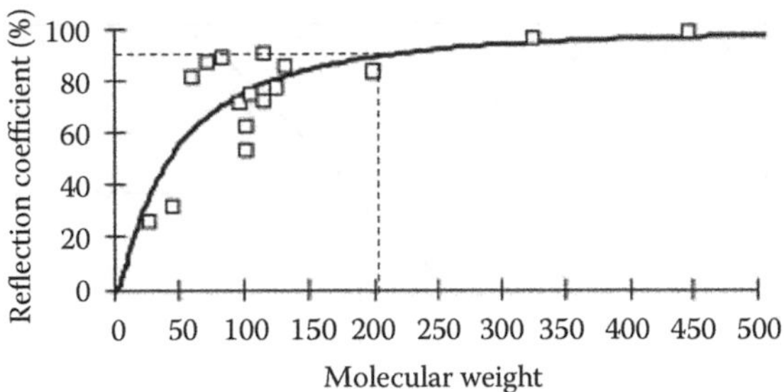

FIGURE 5.12
Rejection rate of NF70 membrane as a function of molecular weight of neutral solutes. (Reprinted from *Water Research*, 36, Van der Bruggen, B., Vandecasteele, C., Modeling of the retention of uncharged molecules with nanofiltration, 1360–1368, Copyright 2002, with permission from Elsevier.)

which the membrane MWCO was determined to be approximately 200 Da (Van der Bruggen and Vandecasteele, 2002). In order to prevent membrane pore blockage and/or pore enlargement during analysis, filtration experiments should be carried out with solutes of progressively higher MW. The permeate samples collected from each type of aqueous solution can then be analyzed and the effective solute rejection, $R(\%)$ of an NF membrane against particular solute can be calculated by

$$R\,(\%) = \left(1 - \frac{C_p}{C_f}\right) \times 100 \tag{5.9}$$

where C_p and C_f are the concentration of the permeate and the feed, respectively. Mostly, the concentration of the sample solution is determined using the total organic carbon (TOC) analyzer and is shown as TOC concentration.

Besides showing MWCO of a membrane, the solute retention measurement can also be directly related to the membrane pore size if the solute rejection is plotted against the Stokes radius of solute tested as shown in Figure 5.13 (Wang and Chung, 2006). In this figure, it was found that the pore radius (in nm) of NF membranes made of PBI in different configurations could be determined using aqueous solutions of different neutral solutes. PBI hollow fiber membrane exhibited pore size of 0.60 nm in radius (equivalent to 525 Da) while PBI flat sheet membrane showed 1.48 nm (equivalent to 3303 Da) in pore radius.

Another NF membrane property—pore size distribution (PSD) can also be established using the results obtained from solute retention measurement. The PSD of a membrane can be determined based on the probability density function as follows:

$$\frac{dR(r_p)}{dr_p} = \frac{1}{r_p \ln \sigma_p \sqrt{2\pi}} \exp\left[-\frac{(\ln r_p - \ln \mu_p)^2}{2(\ln \sigma_p)^2}\right] \tag{5.10}$$

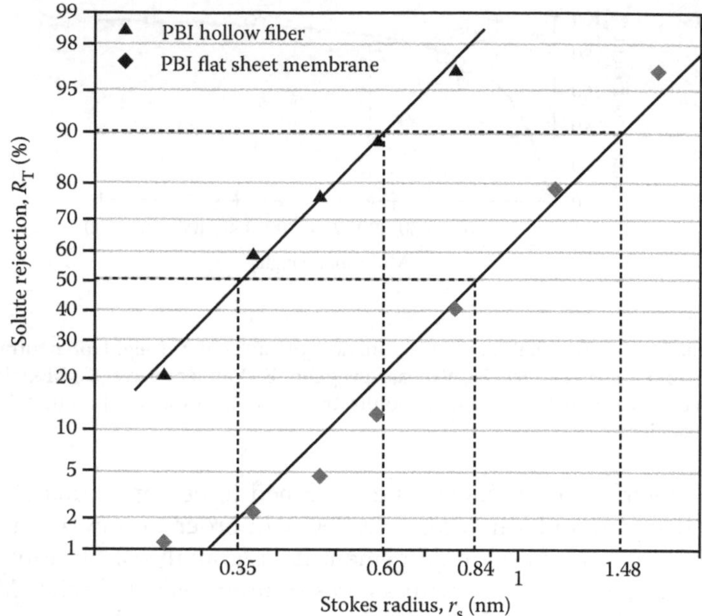

FIGURE 5.13
Effective rejection curves of PBI membranes plotted on the lognormal probability coordinate system (testing conditions: pressure: 20 bar, feed velocity: 0.8 m s^{-1}, feed concentration: 200 mg L^{-1}). (Reprinted from *Journal of Membrane Science*, 281, Wang, K., Chung, T.S., Fabrication of polybenzimidazole (PBI) nanofiltration hollow fiber membranes for removal of chromate, 307–315, Copyright 2006, with permission from Elsevier.)

where r_p is the pore radius, μ_p is the mean pore size, and σ_p is the geometric standard deviation of membrane. Both μ_p and σ_p can be found from the linear relationship of the solute rejection of a membrane as a function of known solute radius (Singh et al., 1998; Ismail and Lau, 2009).

5.4.2.2 Charged Solutes

Many charged solutes are available in the market. However inorganic salts resulting from the neutralization reaction of an acid and a base are commonly used as charged solutes in NF characterization. Salts such as NaCl, Na_2SO_4, $MgSO_4$, and $CaCl_2$ are the most preferable tested solutes used for NF rejection characterization.

For the determination of salt rejection, it is very important to use water of good quality. Deionized water or distilled water with conductivity of less than 5 μS cm^{-1} is recommended as the medium to dissolve ionic compounds. As a rule of thumb, the concentration of salt solution must be in the range of 500–2000 mg L^{-1}. Most preferably, it is reported to be 1000 mg L^{-1}. The rationale for choosing this concentration is mainly because osmotic pressure would become significant when salt concentration in feed solution is very high. Low concentration of salt solution does not create significant osmotic pressure, but surface charge properties of the NF membrane might reject most of the ions which are present in trace amounts due to the Donnan exclusion effect, making the separation efficiency difficult to analyze.

A calibration curve between the conductivity and salt concentration must be established before filtration experiments as the relationship is of importance for determining the salt concentration in the feed and permeate solution measured in conductivity units. Calibration on the conductivity meter can always be performed using commercial salt solutions available at different concentrations, for example, 1000 ppm KCl solution with conductivity value of 1417 μS cm^{-1} at 25°C. The salt rejection, R_{salt}(%) can be evaluated according to the following formula:

$$R_{salt}\ (\%) = \left(1 - \frac{C_p}{C_f}\right) \times 100 \tag{5.11}$$

where C_p is the salt concentration of the permeate solution and C_f is the salt concentration of the feed solution.

The dead end filtration mode is often applied for this characterization, but this technique is not without its drawbacks. It is because the volume of the feed solution is decreasing as a function of filtration time, resulting in an increase in salt concentration (Ong et al., 2012). A high concentration of salt solution can cause significant concentration polarization on the membrane surface, resulting in low permeate flux and possible experimental error. In order to minimize the error, it is highly recommended to use freshly prepared feed solution once the feed concentration is found to increase by more than 10% compared to its original feed concentration. For the cross-flow filtration experiments, the same practice should also be adopted.

5.4.2.3 Dyes

Although standard MWCO characterizations have been developed for aqueous NF systems using test solutes such as PEG, alcohol, and sugar, a standard method for NF in organic solvent systems has yet to be established. This is because the concentration of these organic solutes in the permeate of an SRNF membrane could not be determined by the TOC analyzer as both the solute and solvent contain the carbon (C) element. Therefore, the MWCO of SRNF membranes is typically determined using dyes of different MWs as shown in Table 5.9. This method is quick and effective in evaluating the rejection rate of an SRNF membrane against specific dyes in which the permeate quality can be analyzed by a UV–vis spectrophotometer.

It is known that the interaction mechanisms between solute and water molecules in an aqueous system are less complicated in comparison to the

TABLE 5.9

MW of Several Dyes Together with Its Maximum Absorption Wavelength

Dye	Molecular Weight (g mol^{-1})	Maximum Absorption Wavelength (nm)
Methyl red (MR)	269	496
Crystal violet (CV)	408	586
Reactive orange 16 (RO16)	616	494
Methyl blue (MB)	800	316
Reactive black 5 (RB5)	991	592
Reactive red 120 (RR120)	1470	539

interactions between membrane–solvent–solute in SRNF of an organic system as there are many factors governing the solvent transport in an SRNF membrane. These include (1) physical properties of the solvent such as viscosity, surface tension, dielectric constant, and molecular size, (2) affinity of the solvent (polarity), and (3) surface energy of the membrane. Thus, the use of different solvents for SRNF processes tends to produce different solute rejections following the change in solute size for different solvents, leading to the change of membrane MWCO. Typically, solvents affect the separation performances in two ways: solvent–membrane interaction and solvent–solute interaction. The membrane–solvent interactions have much more impact on the solute separation efficiency compared to the solvent–solute interactions. The solvent–solute interactions, however, are more pronounced when solvents from the same family such as methanol, ethanol, and propanol are compared (Darvishmanesh et al., 2010).

Sani et al. (2014) have studied the rejection profile of an SRNF membrane as a function of dye MW in three different alcohol-based solvents, that is, methanol, ethanol, and isopropanol. Obviously, separation efficiency of the dye varied with the solvent type in which the rejection decreased in the order of isopropanol–ethanol–methanol. This had led to inconsistent MWCO of the SRNF membrane. The results suggested that by changing the solvent properties, the degree of polymer (membrane) swelling might be affected, causing a change in membrane pore dimensions. The degree of membrane swelling can be determined by the difference between solubility parameters of the solvent and membrane. It is generally known that the higher the difference in the solubility parameter, the smaller the degree of membrane swelling. According to Ebert and Cuperus (1999), when a porous membrane swells, the pores become narrower and result in higher rejection. Thus, the higher the degree of swelling, the narrower the pore size of the membrane and the higher the rejection rate. Further analysis shows that the difference between the solubility parameters of the SRNF membrane (made of PPSU) and solvents that is in the order of isopropanol (1.32 MPa$^{1/2}$) ethanol (4.22 MPa$^{1/2}$) methanol (6.92 MPa$^{1/2}$) are consistent with the behavior of dye rejection (Sani et al., 2014).

5.5 Characterization on Membrane Stabilities

5.5.1 Chlorination

A key limitation to interfacially polymerized composite membranes is membrane degradation through contact with chlorine—one of the common disinfectants used in water and wastewater treatment. Chlorine is generally known as the main component attacking the PA structure of composite membrane, forming N-chlorinated amide in the initial step followed by nonreversible ring chlorination, that is, protons on the benzene ring of the PA structure are substituted by Cl via the Orton rearrangement. Because of this, PA membranes prepared must be able to show a certain degree of tolerance against chlorine attack. A chlorine-resistant PA membrane is highly desirable not only for accidental exposure to chlorine, but also for long-term exposure to feed water which uses chlorine to inhibit or prevent growth of microorganisms in water sources. It is reported that the changes in chemical nature of PA upon chlorine exposure is less dependent on feed pH but strongly relies on chlorine concentration and membrane exposure time in feed water (Misdan et al., 2012).

Over the years, efforts have been made to develop novel PA membranes that are highly tolerant to chlorine (Buch et al., 2008). The chlorine tolerance test is usually performed by immersing the PA membrane in an aqueous sodium hypochlorite (NaOCl) solution containing active chlorine at predetermined concentration (100–2500 ppm) for a specified period of time (h). The total amount of membrane exposure to chlorine can be expressed as ppm-h. Practically, the lifespan of a membrane is in the range of several years and the membrane stability tests are, therefore, very time consuming. In order to shorten the testing period, membrane samples are often exposed to chemical components in much higher concentration than expected in practical applications.

To obtain the desired concentration of chlorine solution, one can dilute a commercial NaOCl solution (active chlorine concentration <5%) with pure water. For the test which is conducted for days/weeks, it is recommended to replace the solution everyday with freshly prepared solution as the initial concentration of hypochlorite solution will decrease due to the occurrence of ring chlorination on the PA structure. Also, it is recommended to rinse the tested membrane thoroughly with pure water upon completion of the chlorination test before further use for other assessments.

Water flux and rejection of membrane before and after the NaOCl treatment are recorded to evaluate the membrane chlorine tolerance. A membrane with excellent chlorine tolerance tends to show little changes in water flux and rejection in comparison to a membrane susceptible to chlorine. A PA membrane that is very sensitive to chlorine disinfectant could deteriorate very fast in 1 day even in the case where the chlorine concentration is extremely low (1 ppm). Membrane chlorine tolerance may be further reduced if catalyzing metals such as irons are present in the water source during industrial application. In addition to the filtration experiments, other analyses using IR spectroscopy (surface chemical nature analysis), SEM (surface morphology analysis), and NMR spectroscopy (atom/molecular properties analysis) can also be performed for chlorine stability evaluation (Buch et al., 2008; La et al., 2010).

5.5.2 Thermal

Thermally stable NF membrane is highly desired for some industrial processes which discharge effluents with temperature up to 80–90°C (Marcucci et al., 2003; Schafer et al., 2005). In order to develop a thermally stable PA composite membrane, advanced polymer materials (either for cross-linked PA layer or substrate formation) with good thermal resistance could be utilized (Lau et al., 2012). Generally, the thermal stability test on membrane is conducted by varying the temperature of feed water solution in the range of 20–90°C using a multipurpose immersion coiled heater. Both water flux and rejection rate of a membrane at specified operating temperature are monitored to assess its resistance against thermal attack.

Figure 5.14 shows the example of thermal stability test conducted on a novel melamine–TMC composite membrane and typical PIP–TMC membrane (Han, 2013). The increase in temperature has shown significant improvement on water flux compared to salt rejection, which was constantly maintained. This phenomenon is mainly due to the significant reduction in water viscosity at high operating temperature, causing the water molecules to permeate through the membrane at a faster rate. Comparing these two PA membranes,

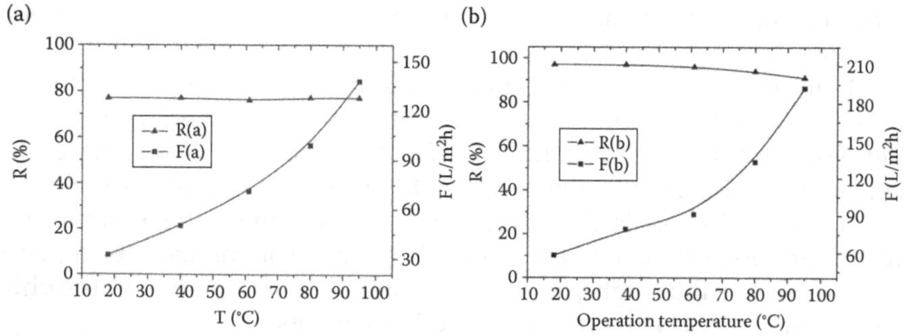

FIGURE 5.14
Thermal stability test on membrane filtration performance, (a) poly(melamine/TMC) composite membrane and (b) PIP–TMC composite membrane (test conditions: 1000 mg L^{-1} Na_2SO_4, 1 MPa, temperature was kept constant for at least 1 h before any data was taken). (Reprinted from *Journal of Membrane Science*, 425–426, Han, R., Formation and characterization of (melamine–TMC) based thin film composite NF membranes for improved thermal and chlorine resistances, 176–181, Copyright 2013, with permission from Elsevier.)

it is fair to say that the melamine–TMC membrane demonstrated greater thermal resistance than the PIP–TMC membrane at high operating temperature with no sign of drop in salt rejection, although its overall salt separation rate was relatively lower. In addition to the filtration experiments, the thermal stability of NF membrane could be further characterized over a wide range of temperature using thermogravimetric analysis (Wu et al., 2010; Rajaeian et al., 2013). Its principal use for composite membrane is to measure the weight change of the membrane as a function of time under a controlled atmosphere. Nevertheless, it must be noted that the weight change of membrane for a temperature less than 100°C is always insignificant, making the analysis on membrane thermal stability very tough.

5.5.3 Solvent

Besides being used in aqueous treatment processes, membrane separation of molecules dissolved in organic liquids has emerged as a new area of application since the last decade. It has shown great potential for many industrial applications. Some of them are reuse of solvents from paints, recovery of sterols and waxes from oils, deacidification of solvent in the vegetable oil refining process, removal of residual triglyceride, glycerol, and unreacted alcohol from biodiesel products, and recovery of APIs from organic liquids.

A general solvent stable composite membrane (including the support) is one that is non-swelling in many solvent categories. The literature has revealed that poly(ethyleneimine)/isophthaloyl dichloride (IPC) is very popular in use in a monomeric system to prepare TFC NF membrane for solvent resistant applications (Korikov et al., 2006; Kosaraju and Sirkar, 2008; Peyravi et al., 2012). Besides SRNF made by the IP technique, the preparation of asymmetric SRNF membrane has also become the focus among membrane scientists in recent years. Some of the polymers that have been previously used for SRNF applications include PPSU, PAN, PI, polypyrrole (PPy), and PEEK.

Similar to the filtration experiment of aqueous salt solution as mentioned earlier, the solvent stability tests can be conducted using either dead-end or cross-flow filtration cell. However, all materials used (e.g., cell body, O-ring/gasket, stirrer, tubing, etc.) for the experiment setup must be tested for stability in solvent prior to the validation process. For experiments that involve highly volatile solvents, it is advised to have a proper liquid sample holder to keep the solution, in order to minimize error during permeability and selectivity determinations.

The stability of an NF membrane against organic solvent can be assessed either by subjecting the membrane sample to a prolonged solvent filtration process or by immersing it in solvent for a specific period of time. Significant changes in the membrane performance, either in permeability or in rejection, are a sign of poor membrane stability against a particular solvent type. Other signs for the influence of organic solvents on the membranes include change in membrane original color, swelling, and shrinking.

5.5.4 Filtration

Although the long-term filtration experiment is quite time consuming, the information obtained from it is of importance to estimate the lifespan of a membrane for practical use. Since most of the lab-scale filtration stability

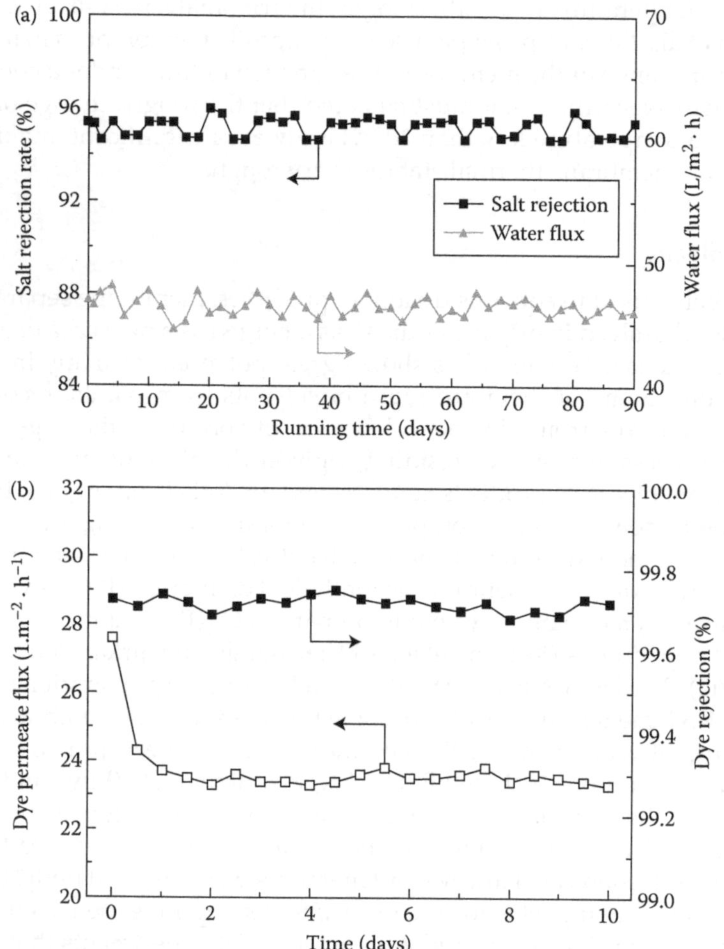

FIGURE 5.15
Long-term stability test on membrane filtration performance, (a) water flux and salt rejection of TFC NF membrane made of PVAm and TMC on PSF substrate (test conditions: 500 mg L^{-1} Na_2SO_4, 0.6 MPa, 25°C, and pH 7) (Reprinted from *Desalination*, 288, Liu, M. et al., Thin-film composite membrane formed by interfacial polymerization of polyvinylamine (PVAm) and trimesoyl chloride (TMC) for nanofiltration, 98–107, Copyright 2012, with permission from Elsevier) and (b) water flux and dye rejection of TFC NF membrane made of PIP and TMC (in the presence of silica) on PES substrate (test conditions: 2000 mg L^{-1} Reactive Brilliant Blue X-BR, 0.6 MPa, and 25°C) (Reprinted from *Desalination*, 301, Hu, D., Xu, Z.-L., Chen, C., Polypiperazine-amide nanofiltration membrane containing silica nanoparticles prepared by interfacial polymerization, 75–81, Copyright 2012, with permission from Elsevier).

tests involve only synthetic solutions with minimal components, the presence of other components in real industrial solution may sometimes play a role in affecting NF membrane performance. To address this problem and get more reliable data, in many cases the construction of membrane pilot plant is necessary so that the membrane can be tested with real industrial solution under a prolonged operating period. Nevertheless, such research articles are rarely found in the open literature.

One of the good examples of the long-term stability test for a TFC NF membrane was reported by Liu et al. (2012b). In this work, the durability of in-house made NF membrane for salt separation was tested over a 90-day period. The changes in water flux and Na_2SO_4 rejection as a function of time is presented in Figure 5.15a. A long-term stability test on NF membrane was also performed using a synthetic dyeing solution as shown in Figure 5.15b (Hu et al., 2012). The main reason for adding dye compounds in the tested solution was to study membrane fouling due to dye adsorption onto the membrane surface. Any changes in NF performances particularly water flux can serve as an indicator on the membrane performance stability. In addition to salt and dye compounds, the assessment on NF membrane fouling propensity can also be performed using a solution containing proteins (Ji et al., 2012), humic acid (Seman et al., 2010), or calcium sulfate (Lin et al., 2006). For instance, the antifouling properties of composite NF membranes were studied using BSA solution by investigating time-dependent water fluxes of membranes during three steps, that is, water flux for 80 min, BSA solution flux for 80 min, and water flux for 80 min after 15 min washing with pure water (Padaki et al., 2013). A membrane with high-water flux recovery (i.e., high reversible) after fouling by BSA solution indicates the excellent antifouling performance of that membrane in reducing adsorption or deposition of protein molecules on the membrane surface during the filtration process.

References

An, Q., Li, F., Ji, Y., Chen, H. 2011. Influence of polyvinyl alcohol on the surface morphology, separation and anti-fouling performance of the composite polyamide nanofiltration membranes. *Journal of Membrane Science*. 367, 158–165.

An, Q., Sun, W., Zhao, Q., Ji, Y., Gao, C. 2013. Study on a novel nanofiltration membrane prepared by interfacial polymerization with zwitterionic amine monomers. *Journal of Membrane Science*. 431, 171–179.

Barona, G.N.B., Choi, M., Jung, B. 2012. High permeate flux of PVA/PSf thin film composite nanofiltration membrane with aluminosilicate single-walled nanotubes. *Journal of Colloid and Interface Science*. 386, 189–197.

Barona, G.N.B., Lim, J., Choi, M., Jung, B. 2013. Interfacial polymerization of polyamide–aluminosilicate SWNT nanocomposite membranes for reverse osmosis. *Desalination*. 325, 138–147.

Buch, P.R., Jagan Mohan, D., Reddy, A.V.R. 2008. Preparation, characterization and chlorine stability of aromatic–cycloaliphatic polyamide thin film composite membranes. *Journal of Membrane Science*. 309, 36–44.

Chen, G., Li, S., Zhang, X., Zhang, S. 2008. Novel thin-film composite membranes with improved water flux from sulfonated cardo poly(arylene ether sulfone) bearing pendant amino groups. *Journal of Membrane Science*. 310, 102–109.

Cheng, X.Q., Liu, Y., Guo, Z., Shao, L. 2015. Nanofiltration membrane achieving dual resistance to fouling and chlorine for "green" separation of antibiotics. *Journal of Membrane Science*. 493, 156–166.

Dalwani, M., Benes, N.E., Bargeman, G., Stamatialis, D., Wessling, M. 2011. Effect of pH on the performance of polyamide/polyacrylonitrile based thin film composite membranes. *Journal of Membrane Science*. 372, 228–238.

Darvishmanesh, S., Degrève, J., Van der Bruggen, B. 2010. Mechanisms of solute rejection in solvent resistant nanofiltration: The effect of solvent on solute rejection. *Physical Chemistry Chemical Physics*. 12, 13333–13342.

Ebert, K., Cuperus, F.P. 1999. Solvent resistant nanofiltration membranes in edible oil processing. *Membrane Technology*. 107, 5–8.

Emadzadeh, D., Lau, W.J., Rahbari-Sisakht, M., Ilbeygi, H., Rana, D., Matsuura, T., Ismail, A.F. 2015. Synthesis, modification and optimization of titanate nanotubes-polyamide thin film nanocomposite (TFN) membrane for forward osmosis (FO) application. *Chemical Engineering Journal*. 281, 243–251.

Fathizadeh, M., Aroujalian, A., Raisi, A. 2011. Effect of added NaX nano-zeolite into polyamide as a top thin layer of membrane on water flux and salt rejection in a reverse osmosis process. *Journal of Membrane Science*. 375, 88–95.

Ghanbari, M., Emadzadeh, D., Lau, W.J., Matsuura, T., Ismail, A.F. 2015. Synthesis and characterization of novel thin film nanocomposite reverse osmosis membranes with improved organic fouling properties for water desalination. *RSC Advances*. 27(5), 21268–21276.

Ghosh, A.K., Hoek, E.M.V. 2009. Impacts of support membrane structure and chemistry on polyamide–polysulfone interfacial composite membranes. *Journal of Membrane Science*. 336, 140–148.

Glater, J., Hong, S., Elimelech, M. 1994. The search for a chlorine-resistant reverse osmosis membrane. *Desalination*. 95, 325–345.

Han, R. 2013. Formation and characterization of (melamine–TMC) based thin film composite NF membranes for improved thermal and chlorine resistances. *Journal of Membrane Science*. 425–426, 176–181.

Hu, D., Xu, Z.-L., Chen, C. 2012. Polypiperazine-amide nanofiltration membrane containing silica nanoparticles prepared by interfacial polymerization. *Desalination*. 301, 75–81.

Ismail, A.F., Lau, W.J. 2009. Theoretical studies on structural and electrical properties of PES/SPEEK blend nanofiltration membrane. *American Institute of Chemical Engineers Journal*. 55, 2081–2093.

Jadav, G.L., Singh, P.S. 2009. Synthesis of novel silica–polyamide nanocomposite membrane with enhanced properties. *Journal of Membrane Science*. 328, 257–267.

Jeong, B.-H., Hoek, E.M.V., Yan, Y., Subramani, A., Huang, X., Hurwitz, G., Ghosh, A.K., Jawor, A. 2007. Interfacial polymerization of thin film nanocomposites: A new concept for reverse osmosis membranes. *Journal of Membrane Science*. 294, 1–7.

Ji, Y., An, Q.F., Zhang Q., Sun W.D., Lee, K.R., Chen, H.L., Gao, C.J. 2012. Novel composite nanofiltration membranes containing zwitterions with high permeate flux and improved anti-fouling performance. *Journal of Membrane Science.* 390–391, 243–253.

Kim, E., Hwang, G., El-din, M.G., Liu, Y. 2012. Development of nanosilver and multiwalled carbon nanotubes thin-film nanocomposite membrane for enhanced water treatment. *Journal of Membrane Science.* 394–395, 37–48.

Kim, S.H.O. 2005. Positron annihilation spectroscopic evidence to demonstrate the flux-enhancement mechanism in membrane. *Environmental Science and Technology.* 39, 1764–1770.

Korikov, A.P., Kosaraju, P.B., Sirkar, K.K. 2006. Interfacially polymerized hydrophilic microporous thin film composite membranes on porous polypropylene hollow fibers and flat films. *Journal of Membrane Science.* 279, 588–600.

Kosaraju, P.B., Sirkar, K.K. 2008. Interfacially polymerized thin film composite membranes on microporous polypropylene supports for solvent-resistant nanofiltration. *Journal of Membrane Science.* 321, 155–161.

Kwak, S., Woo, D. 1999. Use of atomic force microscopy and solid-state NMR spectroscopy to characterize structure-property-performance correlation in high-flux reverse osmosis (RO) membranes. *Journal of Membrane Science.* 158, 143–153.

Kwon, Y., Shih, K., Tang, C., Leckie, J.O. 2012. Adsorption of perfluorinated compounds on thin-film composite polyamide membranes. *Journal of Applied Polymer Science.* 124, 1042–1049.

La, Y., Sooriyakumaran R., Miller, D.C., Fujiwara, M., Terui, Y., Yamanaka, K., McCloskey, B.D., Freeman, B.D., Allen, R.D. 2010. Novel thin film composite membrane containing ionizable hydrophobes: pH-dependent reverse osmosis behavior and improved chlorine resistance. *Journal of Materials Chemistry.* 20, 4615–4620.

Lai, G.S., Lau, W.J., Goh, P.S., Ismail, A.F., Yusof, N., Tan, Y.H. 2016. Graphene oxide incorporated thin film nanocomposite nanofiltration membrane for enhanced salt removal performance. *Desalination.* 387, 14–24.

Lau, W.J., Ismail, A.F., Goh, P.S., Hilal, N., Ooi, B.S. 2015. Characterization methods of thin film composite nanofiltration membranes: A review. *Separation and Purification Reviews.* 44, 135–156.

Lau, W.J., Ismail, A.F., Misdan, N., Kassim, M.A. 2012. A recent progress in thin film composite membrane: A review. *Desalination.* 287, 190–199.

Lee, C.H., Spano, J., Mcgrath, J.E., Cook, J., Freeman, B.D., Wi, S. 2011. Solid-state NMR molecular dynamics characterization of a highly chlorine-resistant disulfonated poly(arylene ether sulfone) random copolymer blended with poly (ethylene glycol) oligomers for reverse osmosis applications. *The Journal of Physical Chemistry B.* 115, 6876–6884.

Lee, H.S., Im, S.J., Kim, J.H., Kim, H.J., Kim, J.P., Min, B.R. 2008. Polyamide thin-film nanofiltration membranes containing TiO_2 nanoparticles. *Desalination.* 219, 48–56.

Li, L., Zhang, S., Zhang, X., Zheng, G. 2008. Polyamide thin film composite membranes prepared from isomeric biphenyl tetraacyl chloride and m-phenylenediamine. *Journal of Membrane Science.* 315, 20–27.

Lin, C.-J., Shirazi, S., Rao, P., Agarwal, S. 2006. Effects of operational parameters on cake formation of $CaSO_4$ in nanofiltration. *Water Research.* 40, 806–816.

Liu, M., Wu, D., Yu, S., Gao, C. 2009. Influence of the polyacyl chloride structure on the reverse osmosis performance, surface properties and chlorine stability of the thin-film composite polyamide membranes. *Journal of Membrane Science*. 326, 205–214.

Liu, M., Yao, G., Cheng, Q., Ma, M., Yu, S., Gao, C. 2012a. Acid stable thin-film composite membrane for nanofiltration prepared from naphthalene-1,3,6-trisulfonylchloride (NTSC) and piperazine (PIP). *Journal of Membrane Science*. 415–416, 122–131.

Liu, M., Zheng, Y., Shuai, S., Zhou, Q., Yu, S., Gao, C. 2012b. Thin-film composite membrane formed by interfacial polymerization of polyvinylamine (PVAm) and trimesoyl chloride (TMC) for nanofiltration. *Desalination*. 288, 98–107.

Liu, Y., Zhang, S., Zhou, Z., Ren, J., Geng Z., Luan, J., Wang, G. 2012c. Novel sulfonated thin-film composite nanofiltration membranes with improved water flux for treatment of dye solutions. *Journal of Membrane Science*. 394–395, 218–229.

Marcucci, M., Ciabatti, I., Matteucci, A., Vernaglione, G. 2003. Membrane technologies applied to textile wastewater treatment. *Annals of the New York Academy of Sciences*. 984, 53–64.

Misdan, N., Lau, W.J., Ismail, A.F. 2012. Seawater reverse osmosis (SWRO) desalination by thin-film composite membrane—Current development, challenges and future prospects. *Desalination*. 287, 228–237.

Misdan, N., Lau, W.J., Ismail, A.F. 2015. Physicochemical characteristics of poly(piperazine-amide) TFC nanofiltration membrane prepared at various reaction times and its relation to the performance. *Journal of Polymer Engineering*. 35(1), 71–78.

Nanda, D., Tung, K.L., Hung, W.S., Lo, C.H., Jean, Y.C., Lee, K.R., Hu, C.C., Lai J.Y. 2011. Characterization of fouled nanofiltration membranes using positron annihilation spectroscopy. *Journal of Membrane Science*. 382, 124–134.

Ng, L.Y., Mohammad, A.W., Leo, C.P., Hilal, N. 2013. Polymeric membranes incorporated with metal/metal oxide nanoparticles: A comprehensive review. *Desalination*. 308, 15–33.

Ong, C.S., Lau, W.J., Ismail A.F. 2012. Treatment of dyeing solution by NF membrane for decolorization and salt reduction and salt reduction. *Desalination and Water Treatment*. 50, 245–253.

Ong, R.C., Chung, T.S. 2012. Fabrication and positron annihilation spectroscopy (PAS) characterization of cellulose triacetate membranes for forward osmosis. *Journal of Membrane Science*. 394–395, 230–240.

Padaki, M., Isloor, A.M., Kumar, R., Ismail, A.F., Matsuura, T. 2013. Synthesis, characterization and desalination study of composite NF membranes of novel poly[(4-aminophenyl)sulfonyl] butanediamide (PASB) and methyalated poly[(4-aminophenyl sulfonyl] butanediamide (mPASB) with polysulfone (PSf). *Journal of Membrane Science*. 428, 489–497.

Peyravi, M., Rahimpour, A., Jahanshahi, M. 2012. Thin film composite membranes with modified polysulfone supports for organic solvent nanofiltration. *Journal of Membrane Science*. 423–424, 225–237.

Rajaeian, B., Rahimpour, A., Tade, M.O., Liu, S. 2013. Fabrication and characterization of polyamide thin film nanocomposite (TFN) nanofiltration membrane impregnated with TiO_2 nanoparticles. *Desalination*. 313, 176–188.

Saha, N.K., Joshi, S.V. 2009. Performance evaluation of thin film composite polyamide nanofiltration membrane with variation in monomer type. *Journal of Membrane Science*. 342, 60–69.

Sani, N.A.A., Lau, W.J., Ismail, A.F. 2014. Influence of polymer concentration in casting solution and solvent–solute–membrane interactions on performance of polyphenylsulfone (PPSU) nanofiltration membrane in alcohol solvents. *Journal of Polymer Engineering.* 34(6), 489–500.

Schaep, J., Vandecasteele, C. 2001. Evaluating the charge of nanofiltration membranes. *Journal of Membrane Science.* 188, 129–136.

Schafer, A.I., Fane, A.G., Waite, T.D. 2005. *Nanofiltration: Principles and Applications.* Oxford: Elsevier.

Seman, M.N.A., Khayet, M., Hilal, N. 2010. Nanofiltration thin-film composite polyester polyethersulfone-based membranes prepared by interfacial polymerization. *Journal of Membrane Science.* 348, 109–116.

Shen, J.N., Yu, C.C., Ruan, H.M., Gao, C.J., Van der Bruggen, B. 2013. Preparation and characterization of thin-film nanocomposite membranes embedded with poly(methyl methacrylate) hydrophobic modified multiwalled carbon nanotubes by interfacial polymerization. *Journal of Membrane Science.* 442, 18–26.

Singh, S., Khulbe, K., Matsuura, T., Ramamurthy, P. 1998. Membrane characterization by solute transport and atomic force microscopy. *Journal of Membrane Science.* 142, 111–127.

Susanto, H., Ulbricht, M. 2009. Characteristics, performance and stability of polyethersulfone ultrafiltration membranes prepared by phase separation method using different macromolecular additives. *Journal of Membrane Science.* 327, 125–135.

Tung, K.-L., Nanda, D., Lee, K., Hung, W., Lo, C., Lai, J. 2009. Characterization of multilayer nanofiltration membranes using positron annihilation spectroscopy. *Journal of Membrane Science.* 343, 147–156.

Van der Bruggen, B., Vandecasteele, C. 2002. Modelling of the retention of uncharged molecules with nanofiltration. *Water Research.* 36, 1360–1368.

Wang, H., Li, L., Zhang, X., Zhang, S. 2010. Polyamide thin-film composite membranes prepared from a novel triamine 3,5-diamino-N-(4-aminophenyl)-benzamide monomer and m-phenylenediamine. *Journal of Membrane Science.* 353, 78–84.

Wang, H., Zhang, Q., Zhang, S. 2011. Positively charged nanofiltration membrane formed by interfacial polymerization of 3,3′,5,5′-biphenyl tetraacyl chloride and piperazine on a poly(acrylonitrile) (PAN) support. *Journal of Membrane Science.* 378, 243–249.

Wang, K., Chung, T.S. 2006. Fabrication of polybenzimidazole (PBI) nanofiltration hollow fiber membranes for removal of chromate. *Journal of Membrane Science.* 281, 307–315.

Wang, L., Li, D., Cheng, L., Zhang, L., Chen, H. 2013. Synthesis of 4-aminobenzoylpiperazine for preparing the thin film composite nanofiltration membrane by interfacial polymerization with TMC. *Separation Science and Technology.* 48, 466–472.

Wu, H., Tang, B., Wu, P. 2010. MWNTs/polyester thin film nanocomposite membrane: an approach to overcome the trade-off effect between permeability and selectivity. *The Journal of Physical Chemistry C.* 114, 16395–16400.

Wu, H., Tang, B., Wu, P. 2013a. Optimizing polyamide thin film composite membrane covalently bonded with modified mesoporous silica nanoparticles. *Journal of Membrane Science.* 428, 341–348.

Wu, H., Tang, B., Wu, P. 2013b. Preparation and characterization of anti-fouling β-cyclodextrin/polyester thin film nanofiltration composite membrane. *Journal of Membrane Science.* 428, 301–308.

Xie, W., Geise, G.M., Freeman, B.D., Lee, H., Byun, G., Mcgrath, J.E. 2012. Polyamide interfacial composite membranes prepared from m-phenylene diamine , trimesoyl chloride and a new disulfonated diamine. *Journal of Membrane Science.* 403–404, 152–161.

Yin, J., Kim, E.-S., Yang, J., Deng, B. 2012. Fabrication of a novel thin-film nanocomposite (TFN) membrane containing MCM-41 silica nanoparticles (NPs) for water purification. *Journal of Membrane Science.* 423–424, 238–246.

Zhang, Q., Wang, H., Zhang, S., Dai, L. 2011. Positively charged nanofiltration membrane based on cardo poly (arylene ether sulfone) with pendant tertiary amine groups. *Journal of Membrane Science.* 375, 191–197.

6

Applications of Nanofiltration Membrane

6.1 Overview of NF Membrane Applications

NF, in concept and operation, is much the same as RO. The key difference is the degree of removal of monovalent ions such as chlorides. Compared to RO membranes which can remove the monovalent ions up to 98%–99%, NF membranes in general show between 50% and 90% rejection. Its level of ion removal varies depending on the material and manufacture of the membrane. For this reason, there are a variety of NF membranes available in the current market for different applications.

Today, NF is mainly applied in drinking water purification process steps such as water softening, removal of sulfates from seawater, chemical oxygen demand (COD) reduction of colored water, and micropollutant elimination. For industrial wastewater treatment processes, NF is used in the textile and paper/pulp industries to reclaim water for reuse through minimum wastewater volume discharge. The use of NF membranes in a nonaqueous medium also holds strong potential in a number of industrial applications since the 1990s. Because of the low energy costs involved in such organic solvent membrane processes, an increasing interest can be observed for applications including solvent recovery in the petrochemical and oleochemical industries as well as separation and purification of valuable products in the pharmaceutical industry.

This chapter does not intend to provide an exhaustive review of all the applications of NF membranes because an extensive review on the membrane applications has been previously published (Schafer et al., 2005). Instead, focus will be placed on the use of NF membranes for several important applications which will be classified according to the nature of the solvent.

6.2 NF of Aqueous Solvent Systems

NF membranes are now an established feature in the water industry for water treatment, water process pretreatment, and wastewater treatment. The following sections provide a brief overview of these applications.

6.2.1 Water Treatment

The NF membrane was essentially developed for groundwater softening where high-levels of calcium, magnesium, and other metal cations were found. The negative charge property of the NF membrane surface is the key element to ensure effective removal of the hard chemicals present in water. In certain cases where total dissolved solid (TDS) reduction is not a primary treatment objective, NF could become an excellent alternative to the lime softening process by eliminating calcium/magnesium ions from naturally hard water while being able to be operated at pressures much lower than that of the RO process. The NF membrane also plays an important role for eliminating color from raw water prior to potable distribution.

For some purposes, NF enables raw water to be treated in a single step to produce permeate of high-quality without any pretreatment. Besides being able to achieve effective removal of hardness, bacteria/virus, dissolved organic matters, and color, this direct NF process brings important environmental benefits, notably by reducing energy requirements and eliminating the needs for dosing chemicals (Koltuniewicz and Drioli, 2008).

However, in many cases, the NF membrane process is integrated with other treatment methods with the purpose of producing water of higher purity and better quality. Figure 6.1 shows the process flow of the *Mery-sur-Oise* water treatment plant for a production capacity of 140,000 m^3 day^{-1} in a suburb of Paris, France. The design of the treatment plant consisted of four important stages, that is, pretreatment (ACTIFLO®), pre-filtration, NF, and post-treatment. The NF membrane (NF200) that was used in this treatment plant was specially developed by DOW FILMTEC™ and has the capability to treat the water from the river *Oise* that contained a high-level of suspended solids as well as pesticides, organic matter, and dissolved organic carbon,

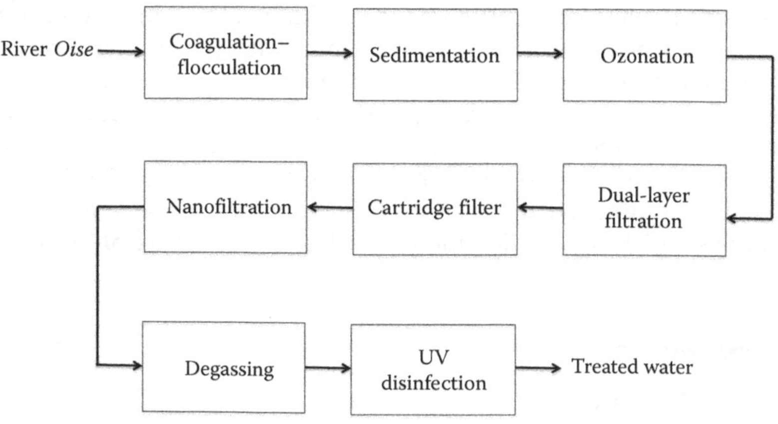

FIGURE 6.1
Integration of NF membrane process in *Mery-sur-Oise* water treatment plant for highly loaded surface water in Paris, France.

in addition to its great variability of temperature (1–25°C) during seasonal changes (Ventresque and Bablon, 1997; Cyna et al., 2002). Using NF200 for the world's largest NF project is proved to be the right choice as it led to no price increase for the residents. Moreover, the organoleptic characteristics of the water have been improved (chlorine taste and calcium concentration are reduced) compared with the conventional plant that operated in parallel.

Another example of using the NF membrane process to produce high-quality drinking water was reported by Anon (2001) in a resort community in Lofsdalen, Sweden. The variable water demand and changes in raw water quality due to seasonal changes prompted the local community to look for a new, safe supply of high-quality drinking water. Prior to the NF filtration process, the raw water from a submerged intake in the level-regulated lake was pretreated by rapid sand filtration and cartridge filtration. The NF255 membrane manufactured by DOW FILMTEC™ (precursor to the NF270 membrane in the current market) was selected to eliminate color and iron while reducing COD. Feed and permeate water analysis as summarized in Table 6.1 indicated that the NF255 membrane demonstrated very high rejections of organic matters, iron, and aluminum. Rejections for divalent and monovalent ions were recorded at 80%–85% and 51%–57%, respectively, which are in good agreement with the separation characteristics of an NF membrane.

In Costa Mesa, California, an NF treatment facility with total permeate production rate of greater than 32,000 m^3 day^{-1} was constructed to expand the current colored water treatment facility (*Membrane Technology*, February 2012). The new treatment system consisted of pretreatment unit, NF system, and post-treatment, and was targeted to remove color in the water source

TABLE 6.1

Performance of NF255 Membrane in Treating the Surface Water of Lake Lofssjön, Sweden

Analysis	Feed (ppm)	Permeate (ppm)	Rejection (%)
Color (Pt)	60–90	2	97–98
COD (Mn)	7–12	<0.2	>98
TOC	13.3	0.05	99.6
UV absorbance at 254 nm	0.58	0.003	99.5
Iron	0.168	<0.003	>98
Aluminum	0.105	<0.005	>95
Calcium	6.15	0.95	84.6
Magnesium	0.84	0.161	80.8
Potassium	0.36	0.154	57.2
Sodium	0.96	0.47	51
Alkalinity, as HCO_3	12	5	58.7
pH	6.9	6.7	–
Conductivity, μS cm^{-1}	43.1	9.81	>77
Silica, mg L^{-1} as Si	2.2	1.93	12.3

taken from Orange Country's groundwater basin prior to potable distribution. The presence of color in water is mainly caused by the buried ancient redwood forest in the area.

The NF process has also been successfully utilized to purify groundwater from a well, producing high-quality water (90–110 m^3 h^{-1}) for a brewery and for soft drink production in a Scandinavian plant (Majamaa et al., 2011). The raw water was pretreated by active carbon filtration followed by a 5 μm safety filter prior to the NF process. NF270 was selected for this large-scale operation as it demonstrated the capacity to reduce feed water alkalinity from 62 mg L^{-1} to an optimal level of 28 mg L^{-1}. Conductivity and total hardness on the other hand were removed by 56.3% and 72.5%, respectively. For the 1.5-year monitoring process, the NF units were consistently operating in a very stable manner with respect to permeate flux, differential pressure, and salt rejection. Most importantly, the end users have been very satisfied with the NF operation and reliability.

More recently, a Canadian membrane filtration company—H_2O Innovation Inc. has completed the commissioning of a drinking water project designed for the city of Delaware, Ohio (*Membrane Technology*, April 2015). In order to provide sufficient potable water for end users, the NF system was integrated with customized pretreatment units to treat water from two different supply sources, that is, from Olentangy River and groundwater. The treated surface water and groundwater with net capacity of 17,034 and 9,464 m^3 day^{-1}, respectively, were ultimately mixed before pumping to the water distribution system.

The NF process as a promising technology for removing arsenic (As) from small/medium-sized water treatment systems also needs to be highlighted. Sato et al. (2002) conducted a survey on the arsenic removal efficiencies of a water treatment plant that consisted of coagulation–sedimentation, inter-chlorination, and manganese sand filtration. The treatment plant had a daily capacity of 65,000 m^3 and was required to treat raw water containing total arsenic and trivalent arsenic (As(III)) in the concentration of 5–50 and 1–10 μg L^{-1}, respectively. The findings of this survey showed that the soluble form of arsenic could not be effectively removed by inter-chlorination and sand filtration, unless a much higher coagulant dosage was used. Optimizing the coagulant dosage for arsenic removal, however, is always not easy to conduct during the treatment process. Besides, this conventional water treatment plant also faced a problem to produce product water with an arsenic level of less than 2 μg L^{-1} when As(III) was present in the raw water. Using the NF membrane (ES-10) manufactured by Nitto Electric Industrial Co., Japan, it was reported that the membrane performance was less susceptible to source water chemical composition. Promising arsenic removal efficiencies (up to 80% for As(III) and 97% for tetravalent arsenic (As(IV)) could be obtained under optimized operating pressure (11 bar) without any chemical additives.

The presence of arsenic and micropollutants in the ground water of the Slavonia region of Croatia, has prompted scientists to consider an effective method—an NF process that could remove both arsenic and pesticides (Kosutic et al., 2005). Water intended for human consumption for this country might contain up to 200 μg L^{-1} of arsenic. And, this region is also known for intensive agriculture that uses various pesticides. Table 6.2 presents the rejection values of NF270 against arsenic and four different types of pesticides. Clearly, the NF membrane showed reasonably high rejection against all the pollutants, except dichlorvos (DDVP). The lowest rejection of DDVP is likely caused by its low *log P* value (the octanol/water partition coefficient) that relates to hydrophobicity. Another case study on the use of NF technology for pesticide removal was documented in Debden Road water works, Saffron Walden, England (Turner and Wittmann, 1998). The process utilizing NF200B from DOW FILMTEC™ was chosen to remove not only the pesticides but also calcium hardness from the water source. During the first year of the plant operation, the pesticide analysis results of the permeate with respect to atrazine, simazine, and chlorotoluron have all been far below drinking water standards (0.1 μg L^{-1}).

The integration of the NF process has several significant advantages compared to conventional water treatment plants for divalent ions, organic matter, and heavy metals removal, but its capital costs might be relatively higher. Nevertheless, it has been industrially proven that the required water quality can be easily attained with the employment of the NF process in conventional treatment processes. This is mainly because conventional treatment processes are very susceptible to changes in feed water characteristics. Surface waters, for example, often change in composition or chemistry owing to seasonal changes or after heavy rainfall. Ground water on the other hand could contain arsenic of different concentrations, depending on geographic locations.

The application of the NF process is also important in the offshore oil and gas industry where removing sulfate from seawater is an important part of preventing scale buildup in wells and reservoirs (*Membrane Technology*, August 2015). Scale deposits are a common problem in water injection as

TABLE 6.2

Rejection Values of Arsenic and Pesticides by NF270 Membrane

		Rejection (%)[a]	
Category	**Component**	**NF270-1**	**NF270-2**
Heavy metal	Arsenic	98.3	95.5
Pesticides	Dichlorvos (DDVP)	40.7	39.3
	Atrazine	81.4	84.8
	Triadimefon	99.8	99.2
	Diazinon	93.1	90.5

[a] NF270 was evaluated twice to yield two different sets of results.

sulfate in the seawater could react with barium and strontium to form barium and strontium sulfate, causing flow problems and subsequent plugging in producing wells. A main benefit of using NF technology (compared to RO) is the fact that the NF elements can be run at lower operating pressure, while still providing high rejection of divalent ions and partial rejection of monovalent ions. The partial demineralization is ideal for maintaining seawater composition and preventing the leaching of minerals from the rocks that the oil adheres to (D'Costa, 2015).

In 1980s, Marathon Oil Co. UK Ltd. worked with FilmTec to develop TFC membranes that could selectively reject sulfate ions in seawater while allowing sodium and chloride ions to pass through the membrane. The result was a product water that was nearly ideal for injection into an offshore reservoir in the North Sea Brae field. A sulfate removal unit (SRU) with the capacity of 40,000 barrels per day was installed and commissioned on the South Brae platform in November 1988 (Davis et al., 1996). A technical report released by GE Power and Water in 2015 revealed that there are over 80 SRUs installed globally with 5–10 new sulfate removal projects being awarded every year (Boczkowski et al., 2015). Typically, an SRU that consists of trains of NF membranes is able to decrease the concentration of incoming sulfate from 2800 to less than 20 ppm. This has led to minimum scale formation and prevented the use of high doses of antiscalants and other chemicals that were normally required in conventional treatment. Boczkowski et al. (2015) emphasized that the long-term performance of SRUs could be further enhanced with the integration of UF, owing to its high efficiency in removing suspended solids and contaminants from seawater prior to entering the NF system.

6.2.2 Wastewater Treatment

In addition to the effective removal of divalent ions, the NF process is also very capable of separating various types of micropollutants with molecular weight that frequently falls in the range of 200–800 g mol^{-1}. The increasing demands on water quality and the need for close to zero effluent discharge have forced industry to consider NF as an alternative treatment option to reduce fresh water consumption.

One of the full-scale NF applications for wastewater treatment was reported in Bangkok, Thailand (*Membrane Technology*, July 2011). The treatment plant consisted of membrane bioreactor (MBR) and the NF unit was designed and constructed by Pentair X-Flow Bv for leachate wastewater separation and purification in Nonthaburi province. It was targeted to treat percolate water that contained high-levels of heavy metals, TOC, biological oxygen demand (BOD), and COD at a production rate of 500 m^3 day^{-1}. A similar hybrid membrane process was also successfully implemented at a landfill site, located near Beacon Hill, Poole in Dorset, UK (Robinson, 2007). By integrating NF with MBR technology, a superior effluent quality with low COD, BOD, and ammonia levels could be produced (see Table 6.3). Clearly, NF was

TABLE 6.3

Typical Characteristics of Raw Leachate, MBR Effluent, and NF Permeate

Parameter	Raw Leachate	MBR Effluent	NF Permeate	NF Rejection (%)
COD (mg L^{-1})	5000	1200	<100	>92
BOD (mg L^{-1})	250	<10	<10	–
Ammonia (mg L^{-1} N)	2000	<2	<1	>50
Temperature (°C)	20	N/A	N/A	–
pH	8	7.1	7.7	–
Total phosphate (mg L^{-1})	15	<5	<1.5	70
Total suspended solid (mg L^{-1})	250	<50	<25	50
Chloride (mg L^{-1})	1400	1200	1200	–
Sulfate (mg L^{-1})	200	200	<10	>95
Conductivity (μS cm^{-1})	16,000	11,000	10,000	9.1
Alkalinity (mg L^{-1})	14,000	200	<50	>75

very effective to reduce the COD value of MBR-treated effluent, ensuring that the permeate could meet the regulatory requirements before discharge to the sewer.

Other references using NF membranes for wastewater treatment are related to textile dyeing effluent. An NF treatment plant designed by PCI-Memtech (Swansea, UK) for Ciba Specialty Chemicals Industries Ltd. in Mahachai, Thailand could process 50 m^3 of highly colored wastewater daily (*Membrane Technology*, July 2002). With the production of high-quality water using NF technology, the volume of wastewater to discard was significantly reduced, making the textile manufacturing more environmentally friendly. On the other hand, commercial NF90 membrane was utilized as a secondary treatment process for a textile factory situated in Valencia, Spain (Gozalvez-Zafrilla et al., 2008). The effluent was pretreated by a biological process to reduce the COD value of raw wastewater from 1500–2000 ppm to less than 200 ppm prior to the NF process. Biological treatment, however, was not effective in the elimination of color and conductivity and NF technology was required to further improve the treated water quality. Figure 6.2 shows the conductivity values of the permeate samples treated by NF process for a period up to 520 h. The inconsistent performance of NF in salt removal could be mainly due to the variations in the volume and characteristics of the wastewater produced. Overall, the NF process was able to achieve at least 90% of salt removal compared with the high conductivity value of the wastewater treated by a biological process (6000–7000 μS cm^{-1}). Further investigation showed that a noticeable reduction of fouling was experienced in the NF system when UF was used as a pretreatment process. Without UF pretreatment, the NF membranes were fouled quickly, leading to comparatively low permeate flux.

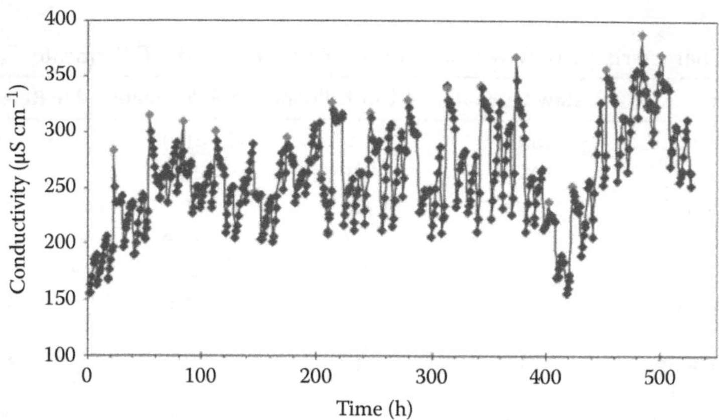

FIGURE 6.2
Conductivity value of permeate treated by NF90-2540 membrane as a function of time. (Reprinted from *Desalination*, 222, Gozalvez-Zafrilla, J.M. et al., Nanofiltration of secondary effluent for wastewater reuse in the textile industry, 272–279, Copyright 2008, with permission from Elsevier.)

An integrated RO–NF process was documented in Karur Amaravathi Textiles, Tamil Nadu, India, aiming to achieve zero discharge of pollutants (Ranganathan et al., 2007). Four RO membrane modules were used, in which the retentate stream of the first module was passed to the second module and so on. Permeate collected from the RO system was then recycled to various textile units for reuse, whereas concentrated streams were diverted to NF for further purification. The concentrated effluent after NF process was finally sent to a solar evaporation pond, leading to no effluent discharge to receiving water bodies. Table 6.4 shows the characteristics of the effluent and the water treated using RO and NF membranes, respectively. In comparison to NF, the RO membrane exhibited much better separation efficiency in the removal of various types of dissolved solutes, including monovalent salts, which are typically retained less by the NF membrane. However, the permeate of the NF has met the minimum requirements for being used to prepare the dye solution of yarn and fabric processes. This case study revealed that the concept of an integrated treatment system in the textile industry is feasible and exhibits huge potential in reducing environmental emissions in addition to reduced energy and material consumption.

The paper pulp industry uses a very great quantity of water in its production processes, a quantity that the industry is striving to reduce, mainly by "closing the water cycle"—a system in which the purification properties of NF have a major role. The advantage of NF in water recovery in this industry is mainly that a cleaner and more consistent water quality can be achieved for recirculation. Some case studies of the NF process used in the pulp and paper industry were reported by Lien and Simonis (1995) and Sutela (2001)

TABLE 6.4

Characteristics of Textile Effluent and Quality of Water Treated by RO and NF Membrane

Parameter	Holding Tank Effluent	Inlet to RO	RO Permeate[b]	RO Retentate	NF Permeate[b]	NF Retentate
pH	8.9	7.8	6.3	8.2	7.6	7.8
Conductivity (mS cm^{-1})	19.3	>20	0.66	>20	>20	>20
TDS (mg L^{-1})	13,770	15,396	364 (97.6)	29,356	28,594 (2.6)	54,734
BOD (mg L^{-1})	213	–[a]	14	600	180 (70.0)	1320
COD (mg L^{-1})	702	–[a]	26	1535	351 (77.1)	3290
Total hardness (mg L^{-1}) as $CaCO_3$	192	163	3 (98.2)	317	107 (66.2)	518
Ca hardness (mg L^{-1}) as $CaCO_3$	115	154	2 (98.7)	307	88 (71.3)	499
Sulfate	1419	1486	38 (97.4)	2254	154 (93.2)	5,644
Chloride	5715	6268	166 (97.4)	12,443	14,931	22,674
Sodium	3900	4146	140 (96.6)	13,000	12,000 (7.7)	22,000
Potassium	84	86	10 (88.4)	230	300	415
Percent sodium	97	98	95	98	98	98
Sodium absorption ratio	122	194	36	317	229	419

[a] Not analyzed.
[b] The value in the bracket represents the separation efficiency (%).

with membrane capacity of 3,000–10,000 and 720 m^3 day^{-1}, respectively. For the NF process to be more effective, it must be operated with suitable pretreatments (Manttari et al., 2010). It is because effluent from paper mills usually contains a large variety of components that are present in different concentrations. These include fibers, colloids, coating pigments, and resins from the broke, sugars, ions, etc. Employing a suitable pretreatment process can ensure a low silt density index feed for the NF process, minimizing flux deterioration caused by fouling. A pilot-scale study has reported that the NF process (using Desal-5 membrane from Osmonics) integrated with biological treatment was probably the best hybrid process to produce treated water of high-quality (Vaisanen et al., 1999). The biological treatment was found to be very effective in digesting foulants such as polysaccharides and lignins that were responsible for NF flux decline in a short operation period. The water quality analysis indicated that the final treated water from this hybrid process was free of sugar, total carbon, pigment, and lipophilics, except for a low level of conductivity.

NF also shows its potential for removing ECs that are normally found in municipal, agricultural, and industrial wastewater sources. Previously, these contaminants were not seriously considered during water quality analysis, but the occurrence of them in water sources is in fact strongly linked to adverse health effects on the human body and other organisms even at extremely low concentration, leading to abnormal sexual differentiation and abnormal development of the nervous and immune systems and other organs. ECs can be generally categorized into four classes, (a) persistent organic pollutants, (b) pesticides, (c) pharmaceutically active compounds (PhACs), and (d) endocrine disrupting chemicals (EDCs).

Reports from lab-scale studies have revealed the effectiveness of NF in removing many of the ECs of concern. This is mainly due to its nanoscale pores and tunable surface charge properties. The utilization of a commercial NF membrane (HL, GE Osmonics) has been found to be able to efficiently remove most of the ECs dissolved in ultrapure water or in municipal secondary effluent (Acero et al., 2010). Of the 11 ECs investigated, the rejection efficiencies of six ECs, that is, metoprolol, antipirine, flumequine, atrazine, hydroxybiphenyl, and diclofenac were promising and remained in the range of 86.1%–100%, 76.7%–88.2%, 80.3%–94.4%, 76.1%–91.3%, 82.7%–96.8%, and 93.3%–98%, respectively, depending on process conditions such as pressure, velocity, temperature, and pH. High rejection against caffeine (85.8%), sulfamethoxazole (97.7%), ketorolac (95.8%), and isoproturon (83.7%) was also possible to achieve when the NF membrane was operated under optimum process conditions.

Furthermore, a study reporting the removal of EDC/PhACs of 52 compounds by NF membrane (ESNA, Hydranautics) has revealed that ECs having less polar, more volatile, and more hydrophobic (group II) compounds such as chrysene, endrin, anthracene, fluorene, naphtalene, etc. were better retained by the membrane than those of having more polar, less volatile, and less hydrophobic (group I) compounds such as triclosan, progesterone, oxybenzone, naproxen, etc. (Yoon et al., 2006). The retention by NF for these compounds is mainly governed by hydrophobic adsorption. A size exclusion mechanism, however, cannot be completely ruled out as it tends to be the dominant factor for these ECs retention once a steady state operation is achieved. Industrially, a membrane treatment plant using tubular NF membrane (AFC30) from PCI-Memtech was constructed to handle a total volume of 1100 m^3 day^{-1} of mixed pharmaceutical effluent, which was previously treated by a combination of biological digestion and dissolved air flotation (*Filtration+Separation*, March 2003). The NF process was designed to further reduce the COD loading of the existing treated effluent, enabling the pharmaceutical company to meet its environmental regulation standards.

Promising results on pesticides removal were also reported when commercial NF70 membrane was used to separate 11 aromatic pesticides with molecular weights ranging from 198 to 286 g mol^{-1} (Chen et al., 2004). Among the pesticides studied, NF70 showed excellent rejection against bentazone (100%), cyanazine (92.2%–96.8%), mecoprop (93.0%–100%), metribuzin (87.5%–97.4%), pirimicarb (100%), and viclozoline (100%), irrespective of the operating pressure and flux recovery rate. Other components, that is, diuron, dinitro-ortho-cresol, metamitron, and simazine could also be removed by NF but at a relatively lower rate (between 46.5% and 92.6% rejection). Similar to other treatment processes, factors such as physiochemical properties of pesticides (e.g., molecular weight, size, hydrophobicity, and polarity) and feed properties (e.g., water pH, solute concentration, presence of dissolved salts, and organic matter) must also be taken into consideration in the mechanisms of pesticide removal, in addition to the membrane characteristics (Plakas and Karabelas, 2012). It is undeniable that NF membrane has proven to be effective in removing many ECs during lab-scale studies, but a great deal of additional data are still needed to fill the gap of knowledge on the long-term stability of its performance.

6.3 NF of Nonaqueous Solvent Systems

Over the last decade, SRNF membranes have emerged as a potential candidate for a wide range of nonaqueous processes either in full-scale industrial process or lab-scale studies. The commonly used materials for the fabrication of NF membranes for aqueous solvent systems are not suitable for organic solutions, although they were the first ones to be used for organic solvents. These hydrophilic polymeric materials (e.g., PA, PSF, and cellulose derivatives) quickly lose their stability when in contact with organic solvents. Because of this reason, special membranes have been developed to provide the same kind of performance as in aqueous systems. The development and commercialization of such NF membranes are in high demand for applications in the pharmaceutical and oil refining industries (Szekely et al., 2014). Up to 2014, 20 different applications have been judged as commercially viable in which three were in the technical stage, eight in the laboratory or pilot phase, and the remaining nine were being looked at as case studies (Lyko, 2014).

6.3.1 Solvent Recovery

The best-known and most successful application of SRNF to date has been in ExxonMobil's MAX-DEWAX process, installed in May 1998 at a refinery in Beaumont, Texas (Bhore et al., 1999; Razdan et al., 2003). This licensing process was co-developed with W.R. Grace & Co. and employed for crude oil dewaxing at a scale of 72,000 barrels per day. It is currently processing around 36,000 barrels per day of feed, consistent with existing permeate demand. Figure 6.3 illustrates the MAX-DEWAX solvent dewaxing process with NF membranes that were made of PI material and packaged as spiral-wound modules. This kind of membrane exhibits chemical integrity and is insoluble in most of the solvents (Gould et al., 1994). It was developed to

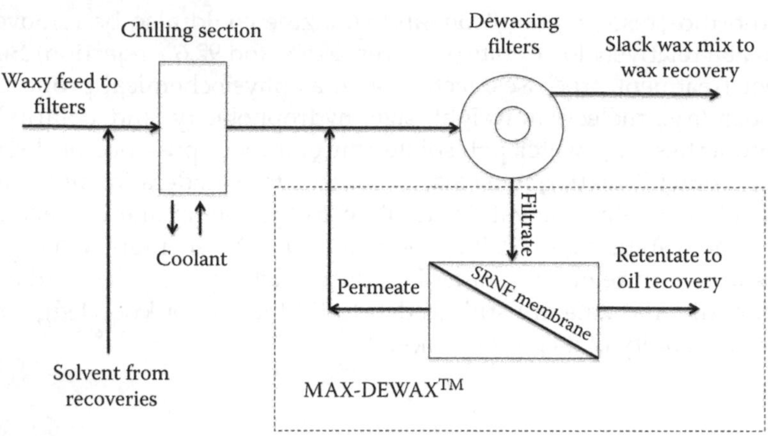

FIGURE 6.3
Schematic of the MAX-DEWAX™ process for cold solvent recovery.

separate solvent from lube oil and hydrocarbon solvent mixtures, cutting the need for energy-intensive distillation, cooling, and refrigeration typically associated with solvent recovery. The preferred solvent is a blend of MEK and toluene. The production data revealed that the NF process combined with ancillary equipment upgrades, increased average base oil production by more than 25% and improved dewaxed oil yields by 3%–5%. Most importantly, the membrane process significantly reduced energy consumption per unit volume of product as much as 20% (Bhore et al., 1999). The capital expenditure of this process (USD 5.5 million) was paid back in less than 1 year by the increase in the net profitability of the lube dewaxing plant. This clearly shows the significant benefits offered by the SRNF process for lube oil production.

Another application of SRNF membrane in the crude oil industry is to recover polar solvent from the naphtenic acid–methanol mixture (Livingston and Osborne, 2002). As organic acids should be removed from crude oil and distilled fractions to avoid corrosion problems particularly in operations where high temperatures (>200°C) are encountered, polar solvent such as methanol is commonly used to extract the acid components. To recover methanol for further extraction, a high energy-intensive distillation process is employed. However, the integration of NF technology in the crude oil deacidification process as illustrated in Figure 6.4 could significantly reduce the energy usage of distillation as only a mixture rich in naphtenic acids (retentate of SRNF) needs to be purified. Depending on the type of SRNF membrane, the solvent flux of 50–100 L m^{-2} h^{-1} and rejection of 76%–87% could be achieved at operating pressure of 30 bar.

Solvent recovery is also important and crucial in the vegetable oil refining process for economic, environmental, and safety reasons. Conventionally, the organic solvent is recovered by distillation followed by a condensation

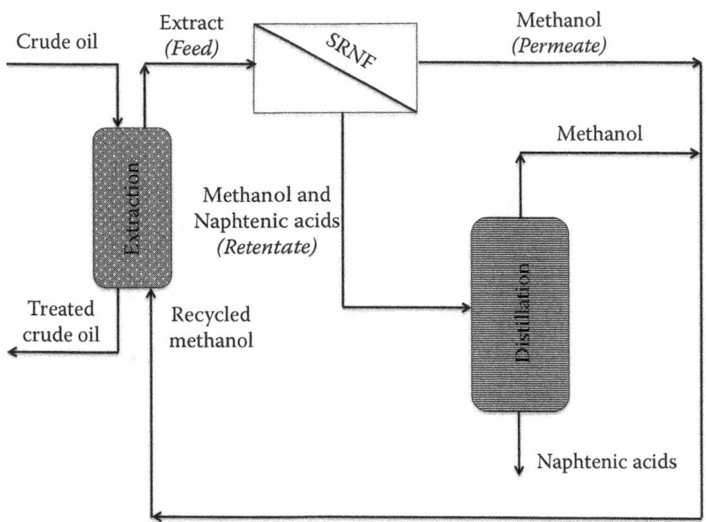

FIGURE 6.4
Process scheme for crude oil deacidification using integrated NF-distillation process.

process. This combined operation, however, is energy intensive and has the disadvantage of handling volatile solvents. A multistage NF process as illustrated in Figure 6.5 has been proposed by Kale et al. (1999) to recover methanol from rice bran oil–methanol mixture. Besides being able to reduce the cost of the new solvent, the proposed system also aimed to minimize the quantity of concentrated free fatty acid (FFA) solution, leading to a smaller footprint of the evaporator. As a single stage NF process has been proven insufficient to completely remove FFA from solvent, a multistage process is therefore required to further purify the solvent before it is recycled for the extractor. Employing solvent for FFA extraction has another significant advantage as it can avoid the use of alkali for neutralization, leading to no soapstock being produced and minimum effluent discharge.

SRNF membrane has also proved potential for recovering ionic liquid that contains reaction products. The process can be carried out by retaining the ionic liquid phase from permeating while allowing organic reaction products to pass through. The introduction of ionic liquids is important to solve the volatility issue of organic solvents as most of the ionic liquids have a close-to-zero vapor pressure at room temperature. This indirectly minimizes the explosion risk of ozone in conventional organic solvents. Moreover, the formation of aerosols is suppressed as the escape of liquid droplets or vapors from the solvent surface is impeded by the high viscosity, high surface tension, and low vapor pressure of ionic liquids. Doorslaer et al. (2010) have assessed the performances of four different types of SRNF membranes and reported that DuraMem® 300 (Evonik Industries) was the best performing membrane in producing ionic liquids of high purity. In the case where

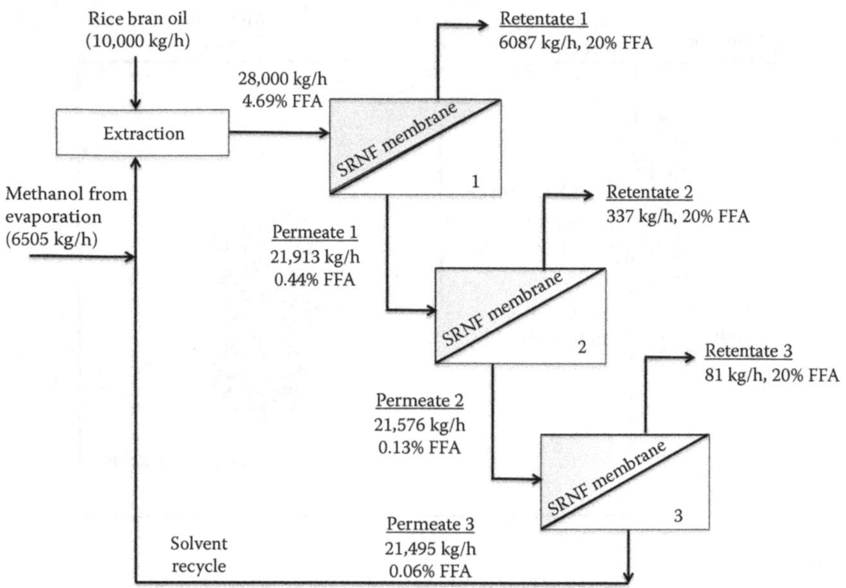

FIGURE 6.5
Conceptual multistage of NF process to recover methanol from rice bran oil–methanol mixture. (With kind permission from Springer Science+Business Media: *Journal of the American Oil Chemists' Society*, Deacidifying rice bran oil by solvent extraction and membrane technology, 76(6), 1999, 723–727, Kale, V., Katikaneni, S.P.R., Cheryan, M.)

[BMPyr][Tf_2N] (MW: 422 g mol^{-1}) was used as ionic liquid, rejection as high as 96% could be achieved after a three-cycle filtration. Similar rejection could even be achieved for the [$MOct_3N$][Tf_2N] of higher MW (649 g mol^{-1}) after only a single filtration step.

6.3.2 Recovery/Purification of Pharmaceutically Active Ingredients and Valuable Catalysts

As more than one-half of the capital investment of pharmaceutical processes is related to separation and purification processes, improvement of the separation efficiencies is of utmost importance for the pharmaceutical industry (Greens et al., 2007). Nevertheless, examples of SRNF membranes for pharmaceutical applications were mainly performed at the laboratory scale using flat sheet membrane and many more are still at the conceptual stage than are in plant use. Some of the crucial limitations of SRNF in the recovery of APIs and catalysts are the low yield of products resulting from insufficient rejection of the product, low solvent permeability, and lack of stability in harsh solvents such as dimethylformamide (DMF) and dichloromethane (Marchetti et al., 2014). The low energy consumption, continuous operation mode, and ease of scaling-up, however, are among the most positive features of the SRNF process compared to other separation processes. Furthermore, the room temperature operation of the NF process could prevent the heat-sensitive APIs from being degraded.

Figure 6.6 illustrates a 400-L pilot SRNF system constructed by Buekenhoudt et al. (2013). It was a real-time monitoring system equipped with pressure, temperature, flow rate, and solvent-level controllers. The system was designed for both ceramic membrane and polymeric membrane with a total membrane surface area of approximately 0.75 and 6 m^2, respectively. The results indicated the huge potential of this pilot system for the API recovery process by concentrating a waste stream containing 1 wt% API to 10 wt%, reducing the amount of waste to be disposed by incineration. The concentrated stream on the other hand could be redirected to existing downstream units for further purification. Another successful implementation of SRNF process is at Janssen Pharmaceutica NV involving an intermediate of a new type of drug (MW = 675 g mol^{-1}) and its oligomeric impurities (MW ≥ 1000 g mol^{-1}) (Sereewatthanawut et al., 2010). Figure 6.7 shows the API purification filtration system that was equipped with the DuraMem® spiral-wound module. It was indicated that 99% of the higher MW oligomeric impurities (i.e., tetramer and higher) have been removed by the DuraMem® membrane, reducing the content of oligomeric impurities in the synthesis solution from 6.8 to 2.4 wt% which is below the allowed limit of 3 wt% oligomeric impurities. With respect to performance stability, the DuraMem® membrane exhibited very stable solvent flux and consistent

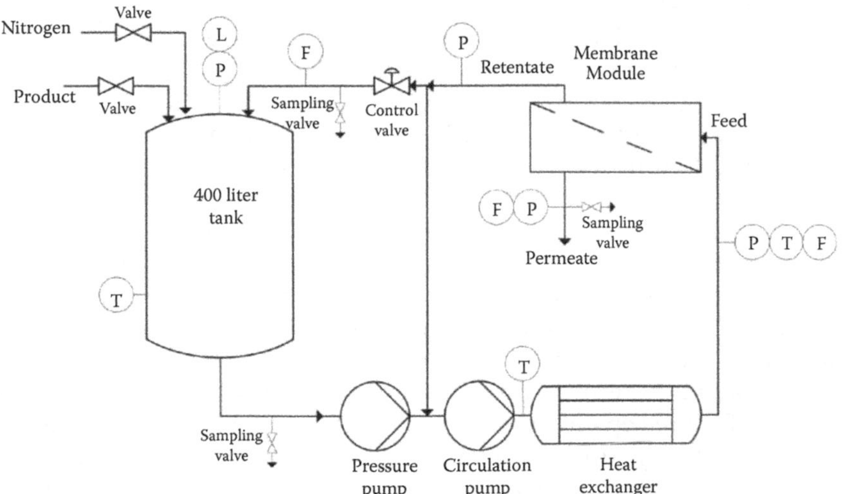

FIGURE 6.6
Pilot-scale SRNF system for recovering API from solvent waste stream. (Buekenhoudt, A. et al.: Solvent based membrane nanofiltration for process intensification. *Chemie Ingenieur Technik*. 2013. 85. 1243–1247. Copyright Wiley-VCH Verlag GmbH & Co. KGaA. Reproduced with permission.)

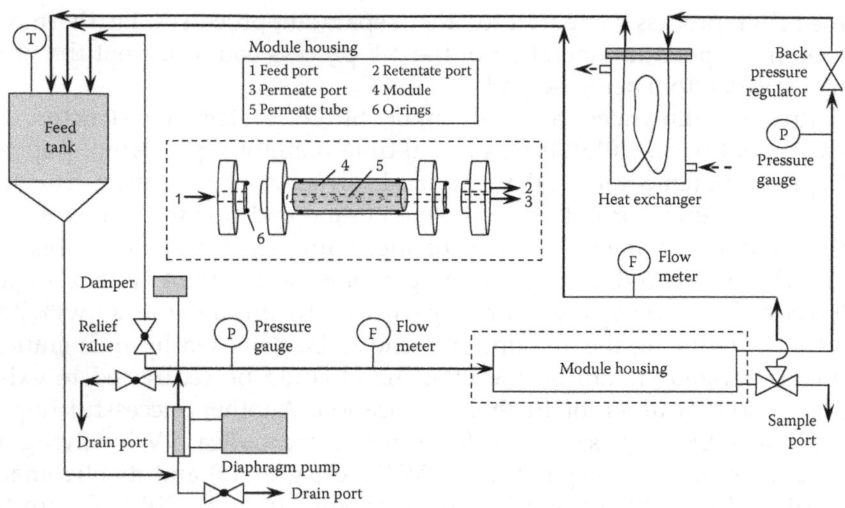

FIGURE 6.7
Pilot plant filtration system for API purification consisting of a 5-L capacity feed tank and membrane housing equipped with DuraMem® spiral-wound module (1.8-in. × 12-in.). (Reprinted with permission from Sereewatthanawut, I. et al. Demonstration of molecular purification in polar aprotic solvents by organic solvent nanofiltration. *Organic Process Research and Development*. 14, 600–611. Copyright 2010 American Chemical Society.)

separation efficiencies when tested with DMF and THF solvent for up to 10 and 120 days, respectively.

A lab-scale study has demonstrated the ability of SRNF membrane to enhance the purity of APIs (in the case of MEK and THF) by removing a wide range of genotoxic impurities (GTIs) (Szekely et al., 2011). Figure 6.8 compares the properties of nine APIs and 11 GTIs with respect to the MW and chemical family. As most of the APIs' MWs are larger than those of the GTIs, separation based on sieve mechanism can take place in which larger APIs are retained while smaller GTIs are allowed to pass through the membrane. As fresh solvent is required to be added to the membrane system to compensate for the solution leaving the system through the permeate (known as diafiltration mode), this purification process is highly solvent intensive. It is therefore recommended to integrate with other processes (e.g., distillation) to recover the solvent. Other works related to the purification of APIs are separation of 1-(5-bromo-fur-2-il)-2-bromo-2-nitroethane from impurities (pyridine acetic anhydride and bromine) (Martínez et al., 2013) and separation of aromatic amine from a dimeric hydrazo impurity (Ormerod et al., 2013b).

Other than using the solvent containing both API and impurities, there are some works reporting the performance of NF with the use of less complicated feed composition. Greens et al. (2007), for example, evaluated the performance of three commercial membranes (StarMem 120, 122, and 228,

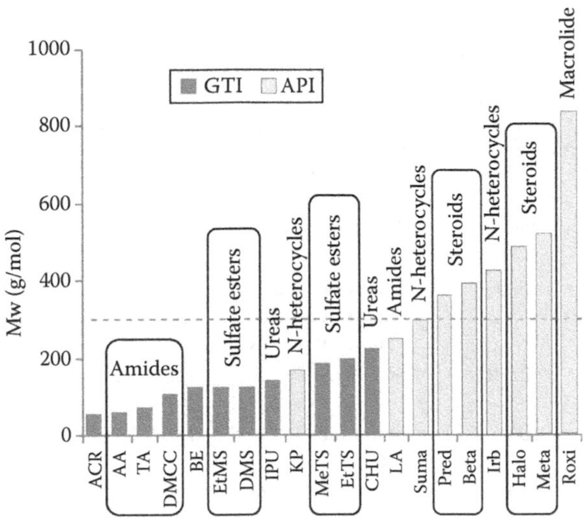

FIGURE 6.8
Comparison between the MW of APIs and GTIs. (ACR: acrolein, AA: acetamide, TA: thioacetamide, DMCC: N-N-dimethylcarbarnoyl-chloride, BE: haloalcohol, EtMS: ethyl mesylate, DMS: dimethylsulfate, IPU: 1,3-diisopropylurea, KP: levetiracetam, MeTS: methyl tosylate, EtTS: ethyl tosylate, CHU: 1,3-dicyclohexylurea, LA: lacosamide, Suma: sumatriptan, Pred: prednisolone, Beta: betamethasone, Irb: irbesartan, Halo: halobetasole, Meta: mometasone, and Roxi: roxithromycin). (Reprinted from *Journal of Membrane Science*, 381, Szekely, G. et al., Organic solvent nanofiltration: A platform for removal of genotoxins from active pharmaceutical ingredients, 21–33, Copyright 2011, with permission from Elsevier.)

Membrane Extraction Technology) in the separation of five APIs of different MW, ranging from 189 to 721 g mol^{-1} in solvents, namely toluene, methanol, and methylene chloride. The specific names and properties APIs were not provided as they were covered by a policy and secrecy agreement with Janssen Pharmaceutica–Johnson & Johnson (Geel, Belgium). The membranes appeared to be resistant against the solvents tested, except in methylene chloride where the top selective layer of membrane was found to dissolve, leading to poor solute rejection. For the other solvents, rejections of greater than 80% were observed for the APIs with a MW of greater than 400 g mol^{-1}. For the API with lower MW of 189 and 313 g mol^{-1}, rejections of 20%–45% and 40%–60%, respectively, were reported, depending on the type of membrane and solvent. Darvishmanesh et al. (2011) on the other hand found that DuraMem 150 from Solsep was a better candidate compared to the StarMem 122 membrane in dealing with organic solvents such as methanol, ethanol, and ethyl acetate. It is because the StarMem membrane experienced swelling on contact with these solvents while the DuraMem membrane maintained a stable performance. Furthermore, the DuraMem membrane exhibited promising separation efficiency (>90%) when tested with APIs namely riluzole (234 g mol^{-1}), atenolol (266 g mol^{-1}), and alprazolam (309 g mol^{-1}).

TABLE 6.5

Applications of SRNF Technology in the Intensification of Catalytic Reactions

Catalyst	MW ($g\ mol^{-1}$)	NF Membrane	Solvent	References
Phosphorus-based organocatalysts	370–430	DuraMem 150/200/300/500	Ethanol, Acetone	Großeheilmann et al. (2015)
Rh-PPh_3-type catalyst	365	PureMem 280, GMT-oNF-2	n-hexane	Schmidt et al. (2014)
Grubbs-Hoveyda and Umicore M catalyst	>500	DuraMem 200, Inopor 0.9 nm TiO_2/1.0 nm TiO_2	Dichloromethane, acetone, toluene	Ormerod et al. (2013a)
Grubbs-Hoveyda II catalyst	627–927	StarMem 122	Toluene	Nasser et al. (2013)
Quinidine-based organocatalyst	1044–1332	DuraMem 300/500	THF	Siew et al. (2013)
$HRh(CO)(PPh_3)_3$	>400	StarMem 122/240	Ethyl acetate	Razak et al. (2013)
POSS enlarged Ru	690	StarMem 288, PuraMem 280	Toluene	Peeva and Livingston (2012)

High value chemicals such as APIs are typically synthesized by multistep reactions that often involve homogeneous organometallic catalysts. Organometallic catalysts recycling can be performed by NF technology as an alternative route. Among the main advantages is that no phase transition or biphasic operation is required by the NF-assisted catalysts recovery. However, in many cases, the catalysts are not large enough to be fully retained by the NF membranes, requiring chemical modification, steric enlargement by means of PEGs, dendrimers, low dispersity poly-(*p*-methylstyrenes), or polyaromatic benzyl bromide supports (Marchetti et al., 2014). The first example for the recycling of a homogenous catalyst by organic solvent membrane was reported by Giggels et al. in 1998 in which a small catalyst was enlarged by attaching it to dendrimers or polymers.

Table 6.5 presents some applications of SRNF technology in the intensification of catalytic reactions in recent years. More information regarding the use of NF membranes in catalytic processes can be found in the review articles authored by Vankelecom (2002) and Marchetti et al. (2014). One example of the industrial applications of SRNF is reported by Evonik Industries for recycling of homogeneous catalysts such as an Rh-based hydroformylation catalyst (Evonik Industries, 2014). Compared to the synthesized product of 200 $g\ mol^{-1}$, the significantly bigger size of the homogeneous catalyst (800 $g\ mol^{-1}$) allowed the NF membrane to recover the catalyst from the product. Reusing the recovered catalyst for subsequent reactions could further reduce the annual budget for catalysts by as much as 80%.

References

Acero, J.L., Benitez, F.J., Teva, F., Leal, A.I. 2010. Retention of emerging micropollutants from UP water and a municipal secondary effluent by ultrafiltration and nanofiltration. *Chemical Engineering Journal*. 163, 264–272.

Anon. 2001. FILMTEC Membranes—Nanofiltration Produces Sparkling Clean Water for Swedish Resort Community. Case History Form No. 609-00379-0503.

Bhore, N.A., Gould, R.M., Jacob, S.M., Staffeld, P.O., McNally, D., Smiley, P.H., Wildemuth, G.R. 1999. New membrane process debottlenecks solvent dewaxing unit. *Oil and Gas Journal*. 97(46), 67–74.

Boczkowski, M., Eriksson, P., Simionato, M. 2015. Water injection and sulfate removal in the offshore oil and gas industry. Technical paper (TP1203EN). GE Power and Water.

Buekenhoudt, A., Bechers, H., Ormerod, D., Bulut, M., Vandezande, P., Vleeshouwers, R. 2013. Solvent based membrane nanofiltration for process intensification. *Chemie Ingenieur Technik*. 85, 1243–1247.

Chen, S.S., Taylor, J.S., Mulford, L.K., Norris, C.D. 2004. Influences of molecular weight, molecular size, flux and recovery for aromatic pesticide removal by nanofiltration membranes. *Desalination*. 160, 103–111.

Cyna, B., Chagneaub, G., Bablon, G., Tanghe, N. 2002. Two years of nanofiltration at the Mery-sur-Oise plant, France. *Desalination*. 147, 69–75.

Darvishmanesh, S., Firoozpour, L., Vanneste, J., Luis, P., Degreve, J. Van der Bruggen, B. 2011. Performance of solvent resistant nanofiltration membranes for purification of residual solvent in pharmaceutical industry: Experiments and simulation. *Green Chemistry*. 13, 3476–3483.

Davis, R., Lomax, I., Plummer, M. 1996. Membranes solve North Sea waterflood sulfate problems. *Oil and Gas Journal*'s website. November, 25.

D'Costa, K. 2015. Seawater sulfate removal with nanofiltration. *Filtration+Separation*. March/April, 22–24.

Doorslaer, C.V., Glas, D., Peeters, A., Odena, A.C., Vankelecom, I., Binnemans, K., Mertens, P., Vos, D.D. 2010. Product recovery from ionic liquids by solvent-resistant nanofiltration: Application to ozonation of acetals and methyl oleate. *Green Chemistry*. 12, 1726–1733.

Evonik Industries. 2014. http://www.design-meets-polymers.com/sites/dc/Downloadcenter/Evonik/Product/DuraMem-PuraMem/brochures/duramem-and-puramem—-general-brochure.pdf.

Filtration+Separation. 2003. Membrane plant for pharmaceutical effluent. *Filtration+Separation*. March, 12.

Giffels, G., Beliczey, J., Felder, M., Kragl, U. 1998. Polymer enlarged oxazaborolidines in a membrane reactor: Enhancing effectivity by retention of the homogeneous catalyst. *Tetrahedron: Asymmetry*. 9(4), 691–696.

Gould, R.M., Heaney, W.F., Nitsch, A.R., Spencer, H.E. 1994. Lubricating oil dewaxing using membrane separation of cold solvent from dewaxed oil and recycle of cold solvent to filter feed. US Patent 5,360,530.

Gozalvez-Zafrilla, J.M., Sanz-Escribano, D., Lora-Garcia, J., Leon Hidalgo, M.C. 2008. Nanofiltration of secondary effluent for wastewater reuse in the textile industry. *Desalination*. 222, 272–279.

Green, J., De Witte, B., Bruggen, B.V.D. 2007. Removal of API's (active pharmaceutical ingredients) from organic solvents by nanofiltration. *Separation Science and Technology*. 42, 2435–2449.

Großeheilman, J., Büttner, H., Kohrt, C., Kragl, U., Werner, T. 2015. Recycling of phosphorus-based organocatalysts by organic solvent nanofiltration. *ACS Sustainable Chemistry and Engineering*. 3, 2817–2822.

Kale, V., Katikaneni, S.P.R., Cheryan, M. 1999. Deacidifying rice bran oil by solvent extraction and membrane technology. *Journal of the American Oil Chemists' Society*. 76(6), 723–727.

Koltuniewicz, A.B., Drioli, E. 2008. *Membrane in Clean Technologies*. Vol. 1, Wiley-VCH, Weinheim, Germany.

Kosutic, K., Furac, L., Sipos, L., Kunst, B. 2005. Removal of arsenic and pesticides from drinking water by nanofiltration membranes. *Separation and Purification Technology*. 42, 137–144.

Lien, L., Simonis, D. 1995. Case histories of two large nanofiltration systems reclaiming effluent from pulp and paper mills for reuse. In: *TAPPI 1995 International Environmental Conference, Atlanta*, Georgia. Book 2, 1023–1027.

Livingston, A.G., Osborne, C.G. 2002. A process for deacidifying crude oil. WO 0250212 A2.

Lyko, H. 2014. Status and perspectives of organophilic nanofiltration. *F&S International Edition*. Issue no. 14/2014, 46–51.

Majamaa, K., Warczok, J., Lehtinen, M. 2011. Recent operational experiences of FILMTEC™ NF270 membrane in Europe. *Water Science and Technology*. 64, 228–232.

Manttari, M., Kallioinen, M., Pihlajamaki, A., Nystrom, M. 2010. Industrial membrane processes in the treatment of process waters and liquors. *Water Science and Technology*. 62, 1653–1660.

Marchetti, P., Jimenez Solomon, M.F., Szekely, G., Livingston, A.G. 2014. Molecular separation with organic solvent nanofiltration: A critical review. *Chemical Reviews*. 14, 10735–10806.

Martínez, M.B., Jullok, N., Rodriguez Negrin, Z., Van der Bruggen, B., Luis, P. 2013. Effect of impurities in the recovery of 1-(5-bromo-fur-2-il)-2-bromo-2-nitroethane using nanofiltration. *Chemical Engineering and Processing: Process Intensification*. 70, 241–249.

Membrane Technology. 2002. Nanofiltration concentrates coloured wastewater and produces potable water. *Membrane Technology*. July, 11–12.

Membrane Technology. 2011. Leachate treatment plant in Bangkok relies on Pentair X-Flow technology. *Membrane Technology*. July, 8.

Membrane Technology. 2012. Biwater AEWT supplies NF system for Costa Mesa water project. *Membrane Technology*. February, 3–4.

Membrane Technology. 2015. H_2O innovation completes its largest integrated UF-NF project to date. *Membrane Technology*. April, 4.

Membrane Technology. 2015. NF membrane removes sea-water sulphate from injection water. *Membrane Technology*. August, 2–3.

Ormerod, D., Bongers, B., Porto-Carrero, W., Giegas, S., Vijt, G., Lefevre, N., Lauwers, D., Brusten, W., Buekenhoudt, A. 2013a. Separation of metathesis catalysts and reaction products in flow reactors using organic solvent nanofiltration. *RSC Advances*. 3, 21501–21510.

Ormerod, D., Sledsens, B., Vercammen, G., Van Gool, D., Linsen, T., Buekenhoudt, A., Bongers, B. 2013b. Demonstration of purification of a pharmaceutical intermediate via organic solvent nanofiltration in the presence of acid. *Separation and Purification Technology*. 115, 158–162.

Peeva, L., Livingston, A.G. 2012. Potential of organic solvent nanofiltration in continuous catalytic reactions. *Procedia Engineering*. 44, 307–309.

Plakas, K.V., Karabelas A.J. 2012. Removal of pesticides from water by NF and RO membranes—A review. *Desalination*. 287, 255–265.

Rabiller-Baudry, M., Nasser, G., Renouard, T., Delaunay, D., Camus, M. 2013. Comparison of two nanofiltration membrane reactors for a model reaction of olefin metathesis achieved in toluene. *Separation and Purification Technology*. 116, 46–60.

Ranganathan, K., Karunagaran, K., Sharma, D.C. 2007. Recycling of wastewaters of textile dyeing industries using advanced treatment technology and cost analysis—Case studies. *Resources, Conservation and Recycling*. 50, 306–318.

Razak, N.S.A., Shaharun, M.S., Mukhtar, H., Taha, M.F. 2013. Separation of hydrido carbonyltris(triphenylphosphine) rhodium (I) catalyst using solvent resistant nanofiltration membrane. *Sains Malaysiana*. 42, 515–520.

Razdan, U., Joshi, S.V., Shah, V.J. 2003. Novel membrane processes for separation of organics. *Current Science*. 85(6), 761–771.

Robinson, T. 2007. Nanotechnology improves landfill leachate quality. *Filtration+ Separation*. November, 38–39.

Sato, Y., Kang, M., Kamei, T., Magara, Y. 2002. Performance of nanofiltration for arsenic removal. *Water Research*. 36, 3371–3377.

Schafer, A.I., Fane, A.G., Waite, T.D. 2005. *Nanofiltration: Principles and Applications*. Elsevier, Oxford, UK.

Schmidt, P., Bednarz, E.L., Lutze, P., Gor´ak, A. 2014. Characterisation of organic solvent nanofiltration membranes in multi-component mixtures: Process design workflow for utilising targeted solvent modifications. *Chemical Engineering Science*. 115, 115–126.

Sereewatthanawut, I., Lim, F.W., Bhole, Y.S., Ormerod, D., Horvath, A., Boam, A.T., Livingston, A.G. 2010. Demonstration of molecular purification in polar aprotic solvents by organic solvent nanofiltration. *Organic Process Research and Development*. 14, 600–611.

Siew, W.E., Ates, C., Merschaert, A., Livingston, A.G. 2013. Efficient and productive asymmetric Michael addition: Development of a highly enantioselective quinidine-based organocatalyst for homogeneous recycling via nanofiltration. *Green Chemistry*. 15, 663–674.

Sutela, T. 2001. Operating experience with membrane technology used for circuit water treatment in different paper mills. In: *COST E14 and PTS-Environmental Technology Symposium*, Grossman, H., Demel, I. (eds), PTS 2001, Munich, Germany, PTS Symposium WU-SY50 108, 11/1–11/17.

Szekely, G., Bandarra, J., Heggie, W., Sellergren, B., Ferreira, F.C. 2011. Organic solvent nanofiltration: A platform for removal of genotoxins from active pharmaceutical ingredients. *Journal of Membrane Science*. 381, 21–33.

Szekely, G., Jimenez-Solomon, M.F., Marchetti, P., Kim, J.F., Livingston, A.G. 2014. Sustainability assessment of organic solvent nanofiltration: From fabrication to application. *Green Chemistry*. 16, 4440–4473.

Turner, A.G., Wittmann, E. 1998. Experience of a nanofiltration plant for pesticide removal. *Membrane Technology*. 104, 7–9.
Vaisanen, P., Huuhilo, T., Puro, L., Nissen, M., Laari, A., Huotari, H., Buchert, J., Suvilampi, J., Nuortila-Jokinen, J., Nystrom, M. 1999. Effect of pretreatments on membrane filtration in paper making process. In: *The Yearbook of Cactus-Technology Program*, VTT Energia, Jyväskylä, Finland, 23–34.
Vankelecom, I.F.J. 2002. Polymeric membranes in catalytic reactors. *Chemical Reviews*. 102, 3779–3810.
Ventresque, C., Bablon, G. 1997. The integrated nanofiltration systemt of Mery-sur-Oise surface water treatment plant (37 mgd). *Desalination*. 113, 263–266.
Yoon, Y., Westerhoff, P., Snyder, S.A., Wert, E.C. 2006. Nanofiltration and ultrafiltration of endocrine disrupting compounds, pharmaceuticals and personal care products. *Journal of Membrane Science*. 270, 88–100.

Index

A

AAPTS, *see* N-[3-(trimethoxysilyl)propyl]ethylenediamine
Active pharmaceutical ingredient, 58, 131, 152–156
Advanced materials in NF membrane, 33–72
  in microporous substrate, 60–72
    polymer/inorganic nanocomposite substrate, 70–72
    polymer/polymer–polymer-based substrate, 60–70
  overview, 33–34
  in PA thin layer, 34–60
    inorganic nanomaterial, 50–60
    monomer, 34–46
    surfactant/additive, 47–50
AFM, *see* Atomic force microscopy
Aluminosilicate SWNTs, 57, 82, 102
Amine monomer, 2, 15–18, 35–40, 42, 49, 85
3-Aminopropyldiethoxymethylsilane, 84
3-Aminopropyltriethoxysilane, 55–56, 85
APDEMS, *see* 3-Aminopropyldiethoxymethylsilane
API, *see* Active pharmaceutical ingredient
Applications (nanofiltration membrane), 139–156
  aqueous solvent systems, 139–149
    wastewater treatment, 144–149
    water treatment, 140–144
  commercial, 6
  nonaqueous solvent systems, 149–156
    pharmaceutically active ingredients and valuable catalysts, recovery/purification of, 152–156
    solvent recovery, 149–152
  overview, 139
APS, *see* 3-Aminopropyltriethoxysilane
Aqueous solvent systems, 139–149
Arsenic (As), 142–143
Asymmetric membrane, 20–26, 47
Atomic force microscopy, 80, 113–114
ATR, *see* Attenuated total reflectance
Attenuated total reflectance, 96–98, 101

B

B-CD, *see* B-Cyclodextrin
B-Cyclodextrin, 107
Biological oxygen demand, 144
Biphenyl tetraacyl chloride, 41
1,4-Bis(3-aminopropyl)piperazine, 38, 42
2,5-Bis(4-amino-2-trifluoromethylphenoxy) Benzenesulfonic acid, 43
4,4′-Bis(4-amino-2-trifluoromethylphenoxy) Biphenyl-4,4′-disulfonic acid, 43
Bisphenol A, 39, 46, 68, 69
BOD, *see* Biological oxygen demand
Bovine serum albumin, 46, 133
BPA, *see* Bisphenol A
BSA, *see* Bovine serum albumin
BTEC, *see* Biphenyl tetraacyl chloride

C

CA, *see* Cellulose acetate
CAGR, *see* Compound annual growth rate
CAIP, *see* Co-solvent assisted interfacial polymerization
Carbon nanotube, 56–57, 71, 89–91, 107, 108
CC, *see* Cyanuric chloride
Cellulose acetate, 1–3
Cellulose nanofiber, 66, 69
Cetyl trimethyl ammonium bromide, 47
Characterization (nanofiltration membrane), 95–133
  chemical properties assessment, instruments/methods for, 96–110

Characterization (nanofiltration membrane) (*Continued*)
attenuated total reflectance-FTIR spectroscopy, 96–98
nuclear magnetic resonance spectroscopy, 109–110
X-ray diffractometry, 107–108
X-ray photoelectron spectroscopy (XPS), 101–107
zeta potential, 98–101
overview, 95–96
permeability, 121–123
physical properties assessment, methods for, 110–121
atomic force microscopy, 113–114
electron microscopy, 110–113
positron annihilation spectroscopy, 114–118
surface contact angle, 119–121
selectivity, 123–128
charged solutes, 126–127
dyes, 127–128
neutral solutes, 123–126
stabilities, 129–133
chlorination, 129–130
filtration, 132–133
solvent, 131
thermal, 130–131
Charged solutes, 123, 126–127
Chemical oxygen demand, 139, 141, 144–145, 148
Chlorination (membrane stability), 45, 96, 129–130
CHMA, *see* 1,3-Cyclohexanebis(methylamine)
CL, *see* Cross-linking
CMCNa, *see* Sodium carboxymethyl cellulose
CN, *see* Cellulose nanofiber
CNT, *see* Carbon nanotube
COD, *see* Chemical oxygen demand
Compound annual growth rate, 5
Copoly(phthalazinone biphenyl ether sulfone), 63, 67
Co-solvent assisted interfacial polymerization, 49
Cross-linking, 17–19, 30, 34, 41, 49–50, 54, 57–58, 66, 69, 101–103, 107
Crude oil deacidification, 151
CTAB, *see* Cetyl trimethyl ammonium bromide
Cyanuric chloride, 43
Cyclohexane, 86
1,3-Cyclohexanebis(methylamine), 39, 44

D

Da, *see* Dalton
Dalton, 123, 125
DAP, *see* Diaminopiperazine
DAPA, *see* 3,5-Diamino-N-(4-aminophenyl) benzamide
DAPP, *see* 1,4-Bis(3-aminopropyl) piperazine
DBES, *see* Doppler broadening energy spectra
Deionized, 122, 126
Deoxyribonucleic acid, 5
DETA, *see* Diethylenetriamine
DI, *see* Deionized
3,5-Diamino-N-(4-aminophenyl) benzamide, 38, 42, 46, 107
Diaminopiperazine, 46
Diethylenetriamine, 41, 118
Dimenthylformanide, 46, 152, 154
Dimenthylsulfoxide, 46, 49
DMF, *see* Dimenthylformanide
DMSO, *see* Dimenthylsulfoxide
DNA, *see* Deoxyribonucleic acid
Donnan steric pore-flow model, 8, 12–13
Doppler broadening energy spectra, 116
DOW FILMTEC™, 144
DSPM, *see* Donnan steric pore-flow model
Dyes, 127–128

E

EC, *see* Emerging contaminant
EDC, *see* Endocrine disrupting chemical
Effluent, 144–148
EGME, *see* Ethylene glycol monomethyl ether
Electron microscopy, 110–113
Emerging contaminant, 6, 147, 148–149
Endocrine disrupting chemical, 148
Ethylene glycol monomethyl ether, 40

F

Fabrication, 79–91
TFC flat sheet membrane, 81–91
alignment of nanotubes/fillers, 89–91
challenges, 81–83
metal alkoxides, use of, 86–87
modified/Novel IP techniques, 87–89
surface modification of nanomaterials, 83–86
TFC hollow fiber membrane, 79–81
challenges, 79–80
innovative approaches, 80–81
6FAPBS, *see* 2,5-Bis(4-amino-2-trifluoromethylphenoxy) Benzenesulfonic acid
6FBABDS, *see* 4,4′-Bis(4-amino-2-trifluoromethylphenoxy) Biphenyl-4,4′-disulfonic acid
FESEM, *see* Field emission scanning electron microscopy
FFA, *see* Free fatty acid
Field emission scanning electron microscopy, 87–88, 110–112, 114, 118
Filtration (membrane stability), 132–133
Flat sheet membrane, TFC, 81–91
FO, *see* Forward osmosis
Forward osmosis, 69–70
Fourier transform infrared, 19, 96–98, 101
Free fatty acid, 151, 152
FTIR, *see* Fourier transform infrared

G

Genotoxic impurities, 154–155
GO, *see* Graphene oxide
Graphene oxide, 52, 55, 71
GTIs, *see* Genotoxic impurities

H

Hagen–Poiseuille model, 9–11
Halloysite nanotube, 82, 108
Hexafluoroalcohol, 37, 45, 109, 110
Hexamethyl phosphoramide, 50
HFA, *see* Hexafluoroalcohol
HMPA, *see* Hexamethyl phosphoramide
HNT, *see* Halloysite nanotube
Hollow fiber membrane, TFC, 79–81
H-OMC, *see* Hydrophilized mesoporous carbon
HPEI, *see* Hyperbranched polyethyleneimine
HTC, *see* 1,3,5-Tricarbonyl chloride
Hydrophilic nanotubes, 82
Hydrophilic surface modifying macromolecule, 48
Hydrophilized mesoporous carbon, 51, 55–57
Hyperbranched polyethyleneimine, 42

I

ICIC, *see* 5-Isocyanato-isophthaloyl chloride
ICP, *see* Internal concentration polarization
IEP, *see* Isoelectric point
Inorganic nanomaterial, 50–60
Interfacial polymerization, 2, 15–20, 33–34, 44–50, 54–60, 66–70, 79–89, 131
Internal concentration polarization, 70
IP, *see* Interfacial polymerization
IPC, *see* Isophthaloyl chloride
IP technique, 15–20, 79–85, 87–89
Irreversible thermodynamic models, 8–9
5-Isocyanato-isophthaloyl chloride, 34, 44
Isoelectric point, 42, 101, 102
Isophthaloyl chloride, 34, 35, 42–43, 45, 58, 131

L

Layer-by-layer assembly method, 15, 26–30
LBL, *see* Layer-by-layer assembly method
Loeb–Sourirajan's dry–wet phase inversion technique, 1
LSMM, *see* Hydrophilic surface modifying macromolecule

M

Manufacturers (of NF membranes), 4
MAX-DEWAX™, 151
MBR, *see* Membrane bioreactor
MEK, *see* Methyl ethyl ketone
Membrane bioreactor, 144, 145
MEOA, *see* Monoethanolamine
Mesoporous silica nanoparticle, 55, 85
Metal-organic framework, 59–60
Methyl ethyl ketone, 58, 150, 154
MF, *see* Microfiltration
Microfiltration, 1
Microporous substrate, 60–72, 97–99
MMPD, *see* M-Phenylenediamine-4-methyl
MOF, *see* Metal-organic framework
Molecular weight, 128, 153–155
  cut-off, 1, 4–5, 5, 123–125, 127–128
Monoethanolamine, 58, 85
Monomer, 2, 15–17, 34–46
MPD, *see* M-Phenyldiamine
M-Phenyldiamine, 2, 18–19, 21, 34, 40–42, 44, 50, 58, 81, 87
M-Phenylenediamine-4-methyl, 37, 44
MSN, *see* Mesoporous silica nanoparticle
Multilayer membrane, 27–30
Multi-walled carbon nanotube, 17, 56–57, 70–71, 82, 84, 90
MW, *see* Molecular weight
MWCNT, *see* Multi-walled carbon nanotube
MWCO, *see* Molecular weight, cut-off

N

Nanofiltration membrane
  applications; *see also* Applications (nanofiltration membrane)
    aqueous solvent systems, 139–149
    commercial, 6
    nonaqueous solvent systems, 149–156
  characterization; *see also* Characterization (nanofiltration membrane)
    chemical properties assessment, instruments/methods for, 96–110
    permeability, 121–123
    physical properties assessment, methods for, 110–121
    selectivity, 123–128
    stabilities, 129–133
  fabrication; *see also* Fabrication
    TFC flat sheet membrane, 81–91
    TFC hollow fiber membrane, 79–81
  global market and research trend of, 4–7
  history, 1–3
  manufacturers, 4
  synthesis, *see* Synthesis (of NF membranes)
  transport mechanisms, 7–13
    Donnan Steric Pore-flow model, 12–13
    Hagen–Poiseuille model, 9–11
    irreversible thermodynamic models, 8–9
    Steric Hindrance Pore (SHP) model, 9–11
    Teorell–Meyer–Sievers model, 11–12
Nanofiltration membrane
  characterization nanofiltration membrane
    physical properties assessment, methods for, 110–121
Neutral solutes, 123–126
NF membrane, *see* Nanofiltration membrane
N-Methyl-2-pyrrolidone, 24
NMP, *see* N-Methyl-2-pyrrolidone
NMR, *see* Nuclear magnetic resonance
Nonaqueous solvent systems, 149–156
Novel IP techniques, 87–89
N-[3-(trimethoxysilyl)propyl]ethylene-diamine, 54, 85, 86
NTSC, *see* 1,3,6-Trisulfonylchloride
Nuclear magnetic resonance spectroscopy, 45, 109–110, 130

O

OCM, *see* Ordered mesoporous carbon
2,2′-OEL, *see* 2,2′-Oxybis-ethylamine
Ordered mesoporous carbon, 51, 55, 57
2,2′-Oxybis-ethylamine, 46

P

PA, *see* Polyamide
PAA, *see* Polyacrylic acid
PAH, *see* Poly(allyl amine) hydrochloride
PA layer, 96–98, 101–103, 108, 111–112, 120
PALS, *see* Positron annihilation lifetime spectroscopy
PAN, *see* Polyacrylonitrile
PANI, *see* Polyaniline
PAS, *see* Positron annihilation spectroscopy
PA thin film (nano) composite membrane, 15–20, 34–60
PBI, *see* Polybenzimidazole
PD, *see* Polydopamine
PDADMAC, *see* Poly(diallyl-dimethyl-ammonium chloride)
PDMS, *see* Polydimethlysiloxane
PE, *see* Polyelectrolyte
PEEK, *see* Poly(ether ether ketone)
PEG, *see* Polyethylene glycol
PEI, *see* Polyetherimide
Permeability, 121–123
Persistent organic pollutant, 148
PES, *see* Polyethersulfone
PhAC, *see* Pharmaceutically active compound
Pharmaceutically active compound, 148
Pharmaceutically active ingredients and valuable catalysts, recovery/purification of, 152–156
Phase inversion technique, 20–26
Phenyl triethoxysilane, 87
PhTES, *see* Phenyl triethoxysilane
PI, *see* Polyimide
PIP, *see* Piperazine
Piperazine, 34, 35, 80, 97
PMMA, *see* Polymethyl methacrylate
Polyacrylic acid, 28
Polyacrylonitrile, 62, 66, 68–69, 98, 131
Poly(allyl amine) hydrochloride, 28
Polyamide, 2, 3, 69
Polyaniline, 68
Polybenzimidazole, 24, 125–126
Poly(diallyl-dimethylammonium chloride), 28–30
Polydimethlysiloxane, 60
Polydopamine, 44–45, 67
Polyelectrolyte, 1, 27
Poly(ether ether ketone), 24, 68, 131
Polyetherimide, 39, 43, 45, 63, 68, 98
Polyethersulfone, 16, 24, 30, 34, 66–68, 70, 97–98
Polyethylene glycol, 44, 98, 124, 127, 156
Polyimide, 60
Polymer/inorganic nanocomposite substrate, 70–72
Polymer/polymer–polymer-based substrate, 60–70
Polymethyl methacrylate, 84
Polyphenylsulfone, 62, 68, 128, 131
Poly(phthalazinone ether amide), 67, 112
Poly(phthalazinone ether nitrile ketone), 64, 67
Poly(phthalazinone ether sulfone ketone), 64, 67–68
Polypropylene, 68
Polypyrrole, 131
Polystyrene, 90
Poly(styrene sulfonate), 28–30
Polysulfone, 16, 24, 25, 28, 60, 68–71, 98, 111, 148
Poly(tetrafluoroethylene), 68
Polythiosemicarbazide, 46
Poly(vinyl alcohol), 57, 83, 107
Poly(vinylamine), 28, 38, 40, 42–43
Polyvinylidene difluoride, 62, 66
Polyvinylpyrrolidone, 98
Poly(vinyl sulfate potassium salt), 28
POP, *see* Persistent organic pollutant
Pore size distribution, 125
Positron annihilation lifetime spectroscopy, 116–117
Positron annihilation spectroscopy, 88, 114, 116–118
PP, *see* Polypropylene
PPBES, *see* Copoly(phthalazinone biphenyl ether sulfone)
PPD, *see* P-Phenylenediamine
PPEA, *see* Poly(phthalazinone ether amide)
PPENK, *see* Poly(phthalazinone ether nitrile ketone)
PPESK, *see* Poly(phthalazinone ether sulfone ketone)

P-Phenylenediamine, 34, 35
PPSU, *see* Polyphenylsulfone
PPy, *see* Polypyrrole
Pre-seeding method, 88
PS, *see* Polystyrene
PSD, *see* Pore size distribution
PSF, *see* Polysulfone
PSS, *see* Poly(styrene sulfonate)
PTFE, *see* Poly(tetrafluoroethylene)
PTSC, *see* Polythiosemicarbazide
Pure water flux, 54, 122–123
Pure water permeability, 122
PVA, *see* Poly(vinyl alcohol)
PVAm, *see* Poly(vinylamine)
PVDF, *see* Polyvinylidene difluoride
PVP, *see* Polyvinylpyrrolidone
PVS, *see* Poly(vinyl sulfate potassium salt)
PWF, *see* Pure water flux
PWP, *see* Pure water permeability

R

Reverse osmosis, 1–3, 34, 70–71, 139–140, 146
Ribonucleic acid, 5
RMS, *see* Root mean square
RNA, *see* Ribonucleic acid
RO, *see* Reverse osmosis
Root mean square, 114

S

SAD, *see* Surface area difference
Scale deposits, 143
Scanning electron microscopy, 87
SDI, *see* Silt density index
SDS, *see* Sodium dodecyl sulfate
Self-assembled PE complex (PEC), 27
SEM, *see* Scanning electron microscopy
Separation spectrum of membrane, 2
SHP, *see* Steric hindrance pore
Silt density index, 147
Single-walled nanotube, 57, 82, 102, 107
SLS, *see* Sodium lauryl sulfate
Sodium carboxymethyl cellulose, 28
Sodium dodecyl sulfate, 47
Sodium lauryl sulfate, 48
Solvent (membrane stability), 131
Solvent recovery, 149–152
Solvent resistant nanofiltration, 46, 58–60, 123, 127–128, 149–156
SPEEK, *see* Sulfonated poly(ether ether) ketone
SPESS, *see* Sulfonated poly(ether sulphide sulfone)
SPPESK, *see* Sulfonated poly(phthalazinone ether sulfone ketone)
SRNF, *see* Solvent resistant nanofiltration
SRNF membrane, 151
SRU, *see* Sulfate removal unit
Steric hindrance pore, 9–11
Sulfate removal unit, 144
Sulfonated poly(ether ether) ketone, 24
Sulfonated poly(ether sulphide sulfone), 60
Sulfonated poly(phthalazinone ether sulfone ketone), 68
Surface area difference, 114
Surface contact angle, 119–121
Surfactant/additive, 47–50
SWNT, *see* Single-walled nanotube
Synthesis (of NF membranes), 15–30
  asymmetric membrane via phase inversion technique, 20–26
  multilayer membrane via layer-by-layer assembly method, 26–30
  overview, 15
  PA thin film (nano) composite membrane via IP technique, 15–20

T

TBP, *see* Tributyl phosphate
TCI, *see* Thionyl chloride
TDS, *see* Total dissolved solid
TEBAB, *see* Triethyl benzyl ammonium bromide
TEBAC, *see* Triethyl benzyl ammonium chloride
TEM, *see* Transmission electron microscopy
TEOA, *see* Triethanolamine
Teorell–Meyer–Sievers model, 11–12
TEPA, *see* Tetraethylenepentamine
TETA, *see* Triethylenetetramine

Tetraethylenepentamine, 41
Tetrahydrofuran, 46, 154
Tetra isopropoxide, 87
TFC membrane, *see* Thin film composite (TFC) membrane
TFN, *see* Thin film nanocomposite
Thermal (membrane stability), 130–131
THF, *see* Tetrahydrofuran
Thin film composite (TFC) membrane, 2–3
  flat sheet membrane, 81–91
  hollow fiber membrane, 79–81
Thin film nanocomposite, 17, 19–20, 33, 51–53, 54–60, 81–89, 108
Thionyl chloride, 58
TMBAB, *see* Trimethyl benzyl ammonium bromide
TMC, *see* Trimesoyl chloride
TOC, *see* Total organic carbon
Total dissolved solid, 140
Total organic carbon, 125, 127, 144
TPP, *see* Triphenyl phosphate
Transmission electron microscopy, 55
Transport models (for NF membranes), 8
Tributyl phosphate, 50
1,3,5-Tricarbonyl chloride, 37, 44
Triethanolamine, 36, 42–43
Triethyl benzyl ammonium bromide, 147
Triethyl benzyl ammonium chloride, 47
Triethylenetetramine, 41, 58, 85
Trimesoyl chloride, 2, 19, 34, 42–50, 55, 58, 81, 85, 107, 109
Trimethyl benzyl ammonium bromide, 47
Triphenyl phosphate, 50
1,3,6-Trisulfonylchloride, 37, 40, 43, 101
TTIP, *see* Tetra isopropoxide

U

Ultrafiltration, 1, 144–145

V

VACNT, *see* Vertically aligned carbon nanotube
Vertically aligned carbon nanotube, 90

W

Wastewater treatment, 144–149
Water treatment, 140–144
Wettability, 119

X

XPS, *see* X-ray photoelectron spectroscope
X-ray diffraction, 107–108
X-ray photoelectron spectroscope, 19, 44, 57, 101–107, 109
XRD, *see* X-ray diffraction

Z

Zeta potential, 98–101

Milton Keynes UK
Ingram Content Group UK Ltd.
UKHW040054071024
449327UK00019B/547